M

SO-ACU-748

Women in Mathematics

The MIT Press
Cambridge, Massachusetts, and London, England

Lynn M. Osen

Women in Mathematics

Thirteenth printing, 1995

Copyright © 1974 by
The Massachusetts Institute of Technology

Printed and bound in the United States of America

Library of Congress Cataloging in Publication Data

Osen, Lynn M
  Women in mathematics.

  CONTENTS: Introduction.—History.—Hypatia, 370-415. [etc.]
  1. Women mathematicians—Biography   I. Title.
QA28.083 1974   510'.92'2 [B]   73-19506
ISBN 0-262-65009-6

This book is dedicated to my son,
Frank S. Osen

Contents

Women in Mathematics grew out of the desire to trace the impact women have had on the development of mathematical thought, to profile the lives of these women, and to explore the social context within which they worked. Beginning with the origin of mathematics in the great civilizations of antiquity and proceeding through the opening decades of the present century, the book deals with many of the notable women whose mathematical accomplishments are best appreciated through an understanding of the societies and the problems that shaped their lives.

In view of the growing interest in the subject of women and their place in history, it is particularly important that we have a more extended statement of their fine legacy in mathematics, and it was to this end that the present volume was prepared. Much has been written about the compelling problems of women in political and economic fields; less attention has centered on women in mathematics and the "hard sciences." It is hoped that this book will in some small measure help to reinforce the idea that women have participated in many of our greatest and most enduring intellectual achievements in these areas, also.

Many women in our present culture value mathematical ignorance as if it were a social grace, and they perceive mathematics as a series of meaningless technical procedures. They discount the role mathematics has played in determining the direction of

Preface

philosophic thought, and they ignore its powerful satisfactions and its aesthetic values, which are equal to those offered by any other branch of knowledge. Yet today more than at any other point in time an appreciation of mathematics is necessary a priori if we are to solve our problems and share in building a better world.

This volume is addressed to the general reader who would like a truer perspective of the women in this discipline, and the book's mathematical statement has been augmented by integrating biography, history, and mathematical content. The general policy has been to include only such mathematics as was useful in placing each subject and her work within the proper context or historical perspective, and one does not need to have a rigorous mathematical interest to appreciate these biographies, though the mathematically oriented reader may find this aspect of the book enlightening and entertaining, also.

It is impossible within the scope of a single volume to do more than give a brief review of the lives of the most outstanding women in mathematics and to trace very lightly their influence, though such a summary is not at all proportionate to the relative interest or importance of these personalities. Nevertheless, an attempt has been made to do something more than catalog the names of these women, and it is hoped that the reader's interest will be piqued by the short histories given here.

In some instances the physical description of a

subject has been included in the text, despite the clear knowledge that male scholars are rarely exposed to such literary treatment. In deference to those whose sensibilities might be offended by this technique, it should be noted that a woman's physical attributes have often determined her life chances, and where such characteristics as beauty (or lack of it), dress, or mien are mentioned in this book, it is in an honest attempt to depict all those aspects of a subject's life that, rightly or wrongly, influenced the course of her career. It is with no condescension or lack of understanding that this rhetorical device has been used here.

It gives me pleasure to acknowledge the many people whose kind advice and assistance contributed to this book: Sylvia Lenhoff of the University of California, Irvine, for her support and encouragement; Dr. Edward Thorp, also of UCI, who was kind enough to read my manuscript; Margaret Kahn, Helen Reister, and the personnel of the Reference Department of UCI Library; and Dr. Gottfried Noether, head of the Department of Statistics at the University of Connecticut, who read the chapter on his aunt and gave me helpful details concerning her life.

I am further indebted to those institutions and individuals who have allowed me to use photographs and illustrations to enhance the book: The National Gallery of Scotland, Scottish National Portrait Gallery of Modern Art; *Scripta Mathematica*; The Bettmann Archive, Inc., New York; and Dag Norberg of Stockholms Universitet.

I should also like to acknowledge my intellectual debt to those mathematical historians whose writings have provided a rich source of information: H. J. Mozans, whose book *Woman in Science*, published early in this century, was a useful guide; Edna Kramer, whose scholarly histories always deal sympathetically with women and whose recent book, *The Nature and Growth of Mathematics*, provided details concerning many of the women of the present century; and Julian L. Coolidge, whose 1951 article in *Scripta Mathematica* helped to spark my own interest in the subject.

Lynn M. Osen

History traditionally bestows only casual recognition on mathematicians, male or female. As Alfred Adler observed in a recent *New Yorker* article, mathematicians come to the realization quite early that "successful research and teaching are the only rewards they will ever receive" (Adler 1972, p. 42). To emphasize his point, Adler challenges his readers to identify more than two of the following: Gauss, Cauchy, Euler, Hilbert, and Riemann. He points out that these are the mathematical equivalents of such greats as Tolstoy, Beethoven, Rembrandt, Darwin, and Freud.

Adler's point is well taken. Only a very few great mathematicians have been lionized with anything like the fervor warranted by their accomplishments; more have faded away into obscurity and mediocrity, while the names of their counterparts in less cerebral disciplines have become extravagantly famous.

But it is interesting to enlarge upon Adler's thesis by examining the role of women in mathematics. If men have had little recognition in this field, women have had less. How many people, indeed, how many mathematicians, recognize the names or contributions of Hypatia, Agnesi, Marquise du Châtelet, Sophie Germain, or Sonya Kovalevsky? Of the last, the German mathematician Kronecker said, "The history of mathematics will speak of her as one of the rarest investigators" (Mozans 1913, p. 164). He was dead wrong. Mathematical history, if it mentions her at all, treats her and her accomplishments in a rather

Introduction

incidental fashion, as it does most other women in this field.

Actually, women today have an exciting and impressive legacy in the development of mathematics. Despite the long tradition of elitism that has surrounded this discipline, a galaxy of brilliant women have made very substantial creative contributions to its development. These contributions are sufficiently worthy of historical note; yet when many of these women are mentioned at all in history texts, it is for their more secular activities, particularly where these have been interesting adjuncts to the life of a famous man.

It is tiresome and counterproductive to argue about the relative merits of male and female mathematicians, for we have no precise method of quantifying or comparing their individual accomplishments. Attempts to rank mathematicians, male or female, into some order or hierarchy of importance have proved to be a very speculative business, one that brings only apocryphal results. There seems to be general agreement that Newton, Gauss, and Archimedes are names of the first order; beyond these, the order is equivocal. For one thing, cross-cultural comparisons do not always yield valid results; each individual mathematician has been a captive of his or her own culture. Nor is it fair to judge mathematicians wholly within the context of their own era, for discoveries and inventions considered trivial in one era have taken on new luster in the light of subsequent developments.

Given these and other impediments, then, there is no acceptable method for assessing or ranking the importance of mathematicians. But even in the absence of such an objective scale, anyone interested in more than casual scholarship notes that women have had a very considerable impact on the development of mathematics. One may also remark that the names of these women are incomparably more obscure than those of their male counterparts. This phenomenon has undoubtedly helped to perpetuate the persistent and pervasive myth that mathematics in its various guises is a male domain.

A recent historian of mathematics has said, and rightly so, "To understand the development of mathematics, we must have a picture of the men who made the science" (Kramer 1955, p. 5). It is just as compelling that we also have a picture of the women who helped to make this science. Not only were their intellectual investments fully as formidable as those of male mathematicians, but they needed an extra dimension of stamina and fortitude to cope with the prejudice against women working in this "male preserve." Society has senselessly confronted these women with a certain fine edge of reproof; only the hardiest personalities could ignore the narrow parochialism of their times or transcend the barbarous conflict of contemporary opinion.

Gauss once wrote a sympathetic letter on this prejudice to Sophie Germain, the great French mathematician. In Gauss's words, "But when a person of the sex, which according to our customs and prejudices, must encounter infinitely more difficulties than men to familiarize herself with these thorny researches, succeeds nevertheless in surmounting these obstacles and penetrating the most obscure parts of them, then without doubt she must have the noblest courage, quite extraordinary talents and a superior genius" (Bell 1937, p. 262).

The lives of these women of "noblest courage, extraordinary talents and a superior genius" amply repay a careful study. It is not only that their scholarly accomplishments form an essential and exciting part of our mathematical history; but their remarkable personal lives also captivate our attention.

Most of these women lived vividly. Their rich personalities and life-styles reflect a variety of characteristics ranging from the wanton to the angelic, from the promiscuous to the pious. Some of them were sensualists; some were ascetics. Some were Pythagoreans; some were bon vivants. Some of them were traditionalists, trying to fit their talents to the ethos of the times; others were flagrant in their refusal to be normatively guided. None of their lives were dull, and indeed some of them were so unusual they have been romanticized in novels, the individuals becoming better known for the more picturesque aspects of their lives than for their scholarship. Even so, their scholarly work concedes nothing in strength, purpose, resolution, and ardor.

Gauss once called mathematics the "queen of the sciences" (Bell 1951), and he went on to observe that she often condescends to render service to astronomy and other natural sciences, but under all circumstances the first place is her due. Gauss did not overstate the role mathematics plays in the service of other branches of science; indeed, his observation is truer today than it was during his own lifetime. It is difficult, therefore, to delimit women's work in mathematics to their contributions as teachers, writers, and researchers, for many more have gone on to use their mathematical training and abilities in other fields.

It is in this more general sense, then, that the term *mathematics* has been used in the present volume, to denote the logical study of number or quantity, shape, space, arrangement, and other related concepts generally accepted as allied with the study of both pure and applied mathematics. In this sense, the temptation has been strong to include chapters on mathematical practitioners who have served as creative forces in other areas of applied science, particularly in the study of astronomy and physics, fields that have traditionally held an especially powerful attraction for women scholars. It is unfair to exclude such women from our notice, but a complete listing of those who have distinguished themselves in these disciplines is outside the province of this volume; however, it should be acknowledged that without their sophisticated comprehension of the mathematical principles available at the time, these women would not have been effective in their endeavors. In some instances, no sharp line distinguishes a woman as an astronomer or as a mathematician. Maria Cunitz, the Silesian who won fame in 1630 for her simplification of Kepler's tables of planetary motion, was an accomplished mathematician whose work helped to reduce the laborious astronomical calculations necessary before her time.

Jeanne Dumée, a Frenchwoman working in astronomy during the same century, used her mathematical training to write an extensive monograph on the theory of Copernicus. Incidentally, she made an elaborate apology for discussing a subject generally not considered a female domain, but she was, according to her words, attempting to prove that women "are not incapable of study, if they wish to make the effort, because between the brain of a woman and that of a man there is no difference" (Mozans 1913, p. 171).

Mme de la Sablière, a contemporary of Jeanne Dumée, also gained more than ordinary fame for her work in astronomy, though she prepared for her career by studying mathematics with the geometer and physicist, Gilles Persone de Roberval. She bore the almost unbearable ridicule heaped on learned Frenchwomen of her time by such writers as Molière and Boileau-Despreaux, who viciously attacked her in his *Contre les Femmes* (published as *Satire X* in 1694). He described Mme de la Sablière as spending her nights watching the planets, a vocation that had ruined her complexion and eyesight, according to the satirist. Unfortunately, she is best known as a target of these barbs rather than for her more serious work. There was also Maria Kirch, a friend of G. W. Leibniz. Some writers (for example Mozans 1913, p. 173) claim that her mathematical calculations enabled her to discover a comet. (Caroline Herschel is usually given the distinction of being the first woman to make such a discovery, and it is interesting to note that no comet was named for Kirch.)

There were many other mathematical practitioners: Mme du Pierry, who authored substantive astronomical tables during the 1700s, and who was the first woman to be professor of astronomy in Paris; Mme Lefrançais de Lalande, who worked out useful methods for finding the time at sea by the altitude of the sun and stars; the Duchesse Louise; Mme Hortense Lepaute, who was a member of the French Academy of Sciences and who worked with Alexis Clairaut to determine the amount of attraction Jupiter and Saturn had on Halley's Comet, a very complex problem for the time. (Incidentally, Clairaut alone is usually given credit for this work.)

American women have also been active in this field: Maria Mitchell, born at Nantucket in 1818, had a talent and love for mathematics that led her into astronomy, and she was the first woman to be made an honorary member of the American Academy of Science. Dorothy Klumpke of San Francisco completed the work begun by Sonya Kovalevsky on the rings of Saturn and was the first woman to be elected a member of the Astronomical Society of France.

During the early part of this century, there was Henrietta Leavitt, whose period-luminosity law was considered an important key to the problem of determining stellar distances. Using the period-luminosity curve set up by Ms. Leavitt, scientists were able to measure the vast stellar distances that were beyond our capabilities with the older parallax method used by Bessell.

Our knowledge of the universe has been enhanced as a result of the work of all these women, and though a closer examination of their lives would be rewarding, it may not be claimed that they were "mathematicians." Thus it is with reluctance that I have excluded their work from a more extensive notice here.

Nor has any attempt been made to include many mathematicians of the present century. Their work is a continuing part of the organic growth of mathematics, and whoever attempts to do full justice to the subject of women in mathematics needs the perspective of history to look back with sympathy and understanding on their work. This is in no way intended to slight the many modern women whose magnificent achievements stand in their own right and need no defense here. It is rather a promise that in the history texts of the future their efforts can be given a more complete acknowledgment.

Even though we cannot apprehend our scriptless, prehistoric past, it seems reasonable to believe that as our antecedents began to evolve the concept of numbers, primitive women developed at a rate at least concomitant with men. If, as philosophers believe, this evolution required untold ages, the shared language and experiential background of the two sexes make it highly unlikely that they developed at differential rates. It was not, perhaps, until the more abstract aspects of mathematics gave a clue to its implicit power that women began to be foreclosed from its intellectual enfranchisement.

A mix of mythology, legend, and fable has blurred our prehistory; it also distorts our view of women's intellectual development during antiquity. However, such anthropologists as Briffault, Mason, Peake, and Keller credit the primitive woman with a relatively high order of creative intelligence, one fully as active and forceful as that of primitive man.

The origins of mathematics and our developing awareness of number and form remain as irretrievable as the beginnings of other branches of human knowledge. Historians can only conjecture about the evolution of mathematics during the centuries prior to 2500 B.C. Apparently by the late Stone Age there were several number systems; by about 3000 B.C. sizable stone buildings were being constructed, and sailing ships had begun to cross the smaller seas. Such activities required a relatively sophisticated

History

knowledge of mathematics, but we have little evidence as to how this knowledge developed.

At about this time, a civilization in which mathematics played a prominent and necessary part had begun to develop in the region around the Tigris and Euphrates rivers within what is now Iraq. Fragments of cuneiform tablets and old calendars dug up by archaeologists in this region suggest that the Babylonians who lived here, perhaps as early as 4700 B.C., were quite competent in mathematics.

The Egyptians also had a calendar as early as 4241 B.C., and the Ahmes papyrus, an Egyptian mathematical text of about 1650 B.C., gives some insight into the recreational activities this culture had devised. Other artifacts indicate that mathematical games were a source of family enjoyment, suggesting that women had some access to the knowledge of the burgeoning mathematics of this culture.

We lack precise details of the gradual development of mathematics during these millennia and cannot say with confidence that women participated in these pursuits. We do know that under Babylonian laws (particularly the Code of Hammurabi) women had certain rights to financial support and could engage in business and own property. Women could be judges, elders, witnesses to documents, and they could also be secretaries. There was a special group of women active in religion to whom frequent reference is made in the Code of Hammurabi, indicating that at least some women were permitted to play a public role in Babylonian society.

In Egypt, also, women could own and inherit property, and they were allowed to engage in trade. Despite the fact that most Egyptian women did not learn to read and write and were not permitted to participate in civil government, a few women were able to become rulers: notably, queen Nitocris in the sixth dynasty around 2000 B.C., Hatshepsut in the eighteenth dynasty, and Tawosret in the nineteenth. At Thebes, daughters of the Tanite kings were installed as "God's Wife of Amon," and these rulers wielded great powers. It is likely such women had access to most of the available knowledge of the times, despite the fact that in the general populace, the role of women was considered to be ancillary.

It is only through examining such subtleties that we can speculate about the extent of women's role in the development of knowledge in these ancient cultures. Certainly no prominent female names in mathematics have come down to us from this time. The same is true of the other great ancient civilizations, such as those of China and India. We can only guess at the role women played in developing a system of mathematics by generalizing from what we know about the status of women in these old cultures, for it was not until the advent of the Hellenic age that the names of a few learned women began to surface in our history.

We do know that the mysteries and power of

mathematics began quite early to come under priestly control, and the mystical characteristics of numbers began to be exploited and emphasized as a valuable religious exegesis. For instance, by plotting the course of the star Sirius, priests were able to predict the annual flooding of the Nile. Such prescience or "magic" powerfully enhanced the priests' ascendancy over less knowledgeable citizens. It was no doubt through such usage that mathematics came to be surrounded by an elitism, remnants of which can be found in our culture today.

The ancient Greeks inherited a legacy of accumulated knowledge and ideas from the older Babylonian and Egyptian cultures—a legacy that the Greeks used well. They were pragmatists; Plato once observed that Greeks customarily borrow the ideas of others, then improve and perfect these ideas, and put them to a useful purpose.

There is some evidence that this is what happened in mathematics, for historians have cited Babylonian tablets that dealt in a sophisticated way with "Pythagorean numbers" a thousand years before Pythagoras' time. There is also evidence that the Chinese had used such numbers as early as 1000 B.C.

One mathematical contribution that is distinctly Greek, however, is the whole notion of proof, deduction, and abstraction. This idea grew out of Greek thought; J. W. N. Sullivan has said of this accomplishment, "The discovery of this unsuspected possibility of the mind was one of the greatest steps forward in the development of human consciousness" (Sullivan 1925, p. 1).

The mathematical formulas in use among the Babylonians had been derived empirically, and they were devised principally for pragmatic purposes, such as surveying and trade. They were not rigorously accurate, nor were they accompanied by proofs. Thales, the teacher of Pythagoras, is credited with being the first man in history to insist on proof rather than intuition in geometry; Pythagoras perpetuated the notion of his teacher. It was his work and teachings that firmly established logical procedures as a Greek tradition.

In some of his work, Pythagoras may have been following the Greek custom of improving on the ideas of others. He had traveled extensively and had spent some time in the Orient learning about the mathematical ideas native to Asia Minor; it is quite likely that he drew on this experience in developing his school at Crotona.

Historians find it difficult to sort fact from legend in the life of Pythagoras, for he was a legendary figure even in his own time. He was born in Samos in 569 B.C., but we know little about his early life. Around 539 B.C., he (along with 300 wealthy supporters in Crotona) established a Dorian colony in southern Italy. This school is said to have created the *science of mathematics*, and the school is of particular importance to our story because Pythagoras was known as the "feminist philosopher."

The Zeitgeist was right for just such a philosophy because the centuries before Pythagoras' time had seen a remarkable increase in the mental vigor of the Greek women. Mozans (1913, p. 7) says of this phenomenon:

Never before nor since did such a wave of feminine genius pass over the fragrant valleys and vine-clad plains of Greece. Never in any other place or time . . . was there a more perfect flowering of female intelligence of the highest order.

Despite this flowering of feminine genius, many of Pythagoras' associates were opposed to sharing the school with women. They remained stubborn and elitist in their attitudes toward education and were inclined to keep the Order's discoveries in arithmetic and geometry secret, not only from women but from the general public as well. Pythagoras, on the other hand, was in favor of disseminating knowledge freely and was willing to teach anyone who was interested. He favored including women in the Order, both as scholars and as teachers; in the end, it was his attitude that prevailed.

One source indicates that there were at least 28 women classified as Pythagoreans who participated in the school. Theano, the beautiful wife and former student of Pythagoras, became a teacher there. She was interested in the study of mathematics, physics, medicine, and child psychology, and she wrote several treatises on these subjects. One of her treatises contained the principle of the "Golden Mean" celebrated as a major contribution of Greek thought to the evolution of social philosophy.

The children of the marriage between Theano and Pythagoras were also involved in the Order, and at least two daughters helped to spread the system of thought that was developed at the school. After the death of Pythagoras, Theano and two of the daughters carried on his work at the central school.

It is impossible to apportion individual credit for work done by the Pythagoreans; whatever was accomplished accrued to the merit of the whole Order, seldom to any one individual. This practice makes it difficult to ascertain the precise role of women, but it is known that the Pythagoreans began as a society of families. Although many modern writers refer to the Order as a "brotherhood," this idea owes more to our social lexicon than to actual fact, for women were serious participants. Indeed, the nature of Pythagorean philosophy required that women be a central part of the Order.

Mathematics took many forms for the Pythagoreans. The premise that *all things are numbers* pervaded the many social elements of the Order: An understanding of mathematics was necessary to the system of philosophy as it related to such subjects as music and harmonics, dancing, songs, and other aesthetic pleasures. Since women shared in these, as well as in many larger social and academic interests, it was necessary that they also share the mathematical philosophy concerning them.

The customs and practices of the Pythagorean philosophy were spread to greater Greece and to Egypt by teachers of both sexes. For over a hundred years, the Pythagorean schools expanded and prospered. They were centers of debate, study, interpretation, and sometimes they were the object of conflict, as when the House of Milo in Crotona was stormed and some fifty members of the Order were murdered. But for many centuries, Greek social thought was infused with the teachings and ideas of the Pythagoreans; so central were the contributions of women to this activity, no conscientious scholar would be content to call the Pythagoreans a "brotherhood."

To trace the derivative effects of the Pythagoreans' feminist activities from the sixth century B.C. through the Hellenic age would take us too far afield. It is relevant to our main theme, however, to mention the influence of the school on such ancient mentors as Plato, Aristotle, and Pericles.

F. M. Cornford writes that Plato's Academy was founded "on lines partly suggested by the Pythagorean societies Plato had seen in South Italy" (Cornford 1953, p. 316). There is other evidence that Plato was influenced by this school, and it is meaningful for the history of social thought that his attitude toward women was a tolerant one.

Plato had a deep appreciation for the intelligence of women and attempted to give them a more equal position of responsibility. He thought education should be compulsory for "all and sundry" and indicated in his writing that

Women ought to share, as far as possible, in education and in other ways with men. For consider: if women do not share in their whole life with men, then they must have some other order of life (Jowett 1892, p. 805).

Plato proposed that both sexes be disciplined in music and gymnastics (here music was used in the Pythagorean sense to include mathematics, literature, astronomy, etc.). He wrote in the *Republic*, "The gifts of nature are alike diffused in both. . . . All the pursuits of men are the pursuits of women" (Jowett 1892, p. 451).

Such attitudes made Plato's school appealing, and women, in defiance of a law that forbade them to attend public meetings, flocked to the groves of his Academy in Athens, where the most important mathematical work of the fourth century B.C. was done.

Both Plato and Socrates generously represented such women as Diotama, Perictione, and Aspasia as being worthy and competent teachers. The latter, an Ionian hetaera, was in many respects one of the most remarkable women Greece ever produced, not only for her influence on Plato but also for her association with Pericles. She convinced the ruler that women should not be denied opportunity for intellectual development. Her political influence secured her a place as one of the most eminent women of her time, but it was her education and genius that won her a place in history. Socrates names her as one of his teachers, and it is believed that she strongly influenced his ideas and those of Plato as expressed in the *Republic* and the *Laws*.

In their extensive and influential writings on social values, both Plato and Socrates took cognizance of women's potential, as did Aspasia, who wrote many of Pericles' speeches. But it should be noted that their opinions regarding women were at sharp variance with those of many of their contemporaries who harbored bias. It is important to note that most women were kept in seclusion and did not participate in the intellectual fervor of the Hellenic age.

During the pre-Christian era, the philosophical schools of Plato and Pythagoras served to create a favorable social climate in which at least some women could pursue an academic career. Because the emphasis on and love of mathematics was so strong in these schools, this tradition persisted long after the Christian era began.

Athenaeus, a Greek writer (ca. A.D. 200), in his *Deipnosophistoe*, mentions a number of women who were superior mathematicians, but precise knowledge of their work in this field is lacking. It is probable that

Hypatia
370–415

there were many women who were well educated in the general science of numbers at this time, judging from the pervasive interest in the subject and the rigor with which women sought an education.

A few Greek women enjoyed comparative freedom in these pursuits, although the class of women known as hetaerae attracted the most public notice. These slave women were usually paramours of the ruling class, although some were freed women or women of free birth; many of them, particularly those from Ionia and Aetolia, strongly impressed themselves on the Greek conscience with their intelligence, wit, and culture. They had keen intellects, and their work in abstract studies made some of them apt students and competent teachers. No doubt the legacy left by these women over the ensuing centuries contributed also to an auspicious social climate within which the formidable genius of Hypatia could flourish in the later part of the fourth century A.D.

Hypatia was the first woman in mathematics of whom we have considerable knowledge, but the story of her life that has come down to us is not a particularly happy one. Despite the good fortune of her legendary talents, her beauty, her long life of hard work, and her celebrated accomplishments in mathematics and astronomy, the story of her eventual martyrdom excites almost the same sympathies as a classic Greek tragedy. Although nearly a thousand years separated her from the time of Aspasia, in many ways Hypatia was also a true daughter of Greece.

Hypatia was born around A.D. 370, and her father, Theon, was a distinguished professor of mathematics at the University of Alexandria. He later became the director of the University, and Hypatia's early life was spent in close contact there with the institute called the *Museum*.

We know little about Hypatia's mother, but the family situation must have been a fortunate one, for Theon was determined to produce a perfect human being. As Elbert Hubbard (1908, p. 83) remarked, ". . . whether his charts, theorems and formulas made up a complete law of eugenics, or whether it was dumb luck, this we know: he nearly succeeded."

From her earliest years Hypatia was immersed in an atmosphere of learning, questioning, and exploration. Alexandria was the greatest seat of learning in the world, a cosmopolitan center where scholars from all the civilized countries gathered to exchange ideas. As Theon's daughter, Hypatia was a part of this stimulating and challenging environment. In addition, she received a very thorough formal training in arts, literature, science, and philosophy.

Theon was his daughter's tutor, teacher, and playmate; his own strong love of the beauty and logic of mathematics was contagious. We know that he was influential in this part of Hypatia's intellectual development, which was eventually to eclipse his own.

At the time, mathematics was used mainly for calculating such obscure problems as the locus of a given soul born under a certain planet. It was thought that mathematical calculations could determine precisely where such a soul would be on a future date. Astronomy and astrology were considered one science, and mathematics was a bond between this science and religion.

These disciplines were a part of Hypatia's early training, and, in addition, Theon introduced her to all the systems of religion known to that part of the civilized world. He had a rare talent as a teacher, and he was determined to transmit to Hypatia not only the accumulated fund of knowledge but the discrimination needed to assimilate and build upon this fund. Toward this end, he was particularly concerned that she be discriminate about religion and that no rigid belief take possession of her life to the exclusion of new truths. "All formal dogmatic religions are fallacious and must never be accepted by self-respecting persons as final," he told her. "Reserve your right to think, for even to think wrongly is better than not to think at all" (Hubbard 1908, p. 82).

Theon also established a regimen of physical training to ensure that Hypatia's healthy body would match her formidable, swift, well-trained mind. He devised a series of gentle calisthenics that she practiced regularly; she was taught to row, swim, ride horseback, and climb mountains, and a part of each day was set aside for such exercise.

To the Romans the art of the rhetor, or orator, was one of the most consequential of the social graces; the ability to impress others by one's personal presence was indeed a most extraordinary gift. As part of the preparation for becoming the "perfect human being" that Theon had determined she should be, Hypatia was given formal training in speech, and there were lessons in rhetoric, the power of words, the power of hypnotic suggestion, the proper use of her voice, and the gentle tones considered pleasing. Theon structured her life minutely and precisely, leaving little to chance or circumstance, but he was not content to produce such a powerful personality without giving her an understanding of her responsibility to others. He cautioned her about the vulnerability of the permeable, impressionable mind of the young, and he warned her against using the cosmetic effect of rhetoric and pretense to influence or manipulate others. His training urged her toward becoming a sensitive, gifted, and eloquent teacher, and these qualities are reflected in her writing:

Fables should be taught as fables, myths as myths, and miracles as poetic fancies. To teach superstitions as truths is a most terrible thing. The child mind accepts and believes them, and only through great pain and perhaps tragedy can he be in after years relieved of them. In fact men will fight for a superstition quite as quickly as for a living truth— often more so, since a superstition is so intangible you cannot get at it to refute it, but truth is a point of view, and so is changeable (Hubbard 1908, p. 84).

As a further part of her education, Hypatia traveled abroad and was treated as royalty wherever she went. Some accounts say that Hypatia's travels extended over a period of 10 years; others say she spent only a year or so in travel. It is probable that her trips extended over a long period of time and were not continuous, but it is known that for a while she was a student in Athens at the school conducted by Plutarch the Younger and his daughter Asclepigenia. It was here that her fame as a mathematician became established, and upon her return to Alexandria, the magistrates invited her to teach mathematics and philosophy at the university. She accepted this invitation and spent the last part of her life teaching out of the chair where Ammonius, Hierocles, and other celebrated scholars had taught.

She was a popular teacher; Socrates, the historian, wrote that her home, as well as her lecture room, was frequented by the most unrelenting scholars of the day and was, along with the library and the museum, one of the most compelling intellectual centers in that city of great learning. She was considered an oracle, and enthusiastic young students from Europe, Asia, and Africa came to hear her lecture on the *Arithmetica* of Diophantus, the techniques Diophantus had developed, his solutions of indeterminate problems of various types, and the symbolism he had devised. Her lectures sparkled with her own mathematical ingenuity, for she loved mathematics for its own sake, for the pure and exquisite delight it yielded her inquisitive mind.

Hypatia was the author of several treatises on mathematics. Suidas, the late-tenth-century lexicographer of Greek writings, lists several titles attributed to her, but unfortunately these have not come down to us intact. Most were destroyed along with the Ptolemaic libraries in Alexandria or when the temple of Serapis was sacked by a mob, and only fragments of her work remain. A portion of her original treatise *On the Astronomical Canon of Diophantus* was found during the fifteenth century in the Vatican library; it was most likely taken there after Constantinople had fallen to the Turks.

Diophantine algebra dealt with first-degree and quadratic equations; the commentaries by Hypatia include some alternative solutions and a number of new problems that she originated. Some scholars consider these to have been in Diophantus' original text, but Heath (1964, p. 14) attributes them to Hypatia.

In addition to this work, she also wrote *On the Conics of Apollonius*, popularizing his text. It is interesting to note that, with the close of the Greek period, interest in conic sections waned, and after Hypatia, these curves were largely neglected by mathematicians until the first half of the seventeenth century.

Hypatia also wrote commentaries on the *Almagest*,

the astronomical canon of Ptolemy's that contained his numerous observations of the stars. In addition, she coauthored (with her father) at least one treatise on Euclid. Most of these works were prepared as textbooks for her students. As was the case with her commentaries on *Conics*, no further progress was made in mathematical science as taught by Hypatia until the work of Descartes, Newton, and Leibniz many centuries later.

Among Hypatia's most distinguished pupils was the eminent philosopher Synesius of Cyrene, who was later to become the wealthy and influential Bishop of Ptolemais. His letters asking for scientific advice have furnished us with one of the richest sources of information concerning Hypatia and her works, and they indicate how keenly he valued his intellectual association with her (see, for example, Hale 1860, p. 111).

References are found in Synesius' letters crediting Hypatia with the invention of an *astrolabe* and a *planesphere*, both devices designed for studying astronomy. His letters also credit her with the invention of an apparatus for distilling water, one for measuring the level of water, and a third for determining the specific gravity of liquids. This latter device was called an *aerometer* or *hydroscope*.

Hypatia's contemporaries wrote almost lyrically about her great genius. Socrates, Nicephorus, and Philostorgius, all ecclesiastical historians of a persuasion different from that of Hypatia, nevertheless were generous in their praise of her characteristics and learning. Her popularity was wide and genuine, and it is said that she had several offers of marriage from princes and philosophers, but to these proposals she answered that she was "wedded to the truth." This pretty speech was no doubt more an evasion than a verity; it is more likely that she simply never met a suitor whose mind and philosophy matched her own. Although she never married, she did have love affairs, and various imaginary romances have been credited to her.*

Her renown as a philosopher was as great as her fame as a mathematician, and legend has it that letters addressed to "The Muse" or "The Philosopher" were delivered to her without question. She belonged to a school of Greek thought that was called neo-Platonic: the scientific rationalism of this school ran counter to the doctrinaire beliefs of the dominant Christian religion, seriously threatening the Christian leaders. These pietists considered Hypatia's philosophy heretical, and when Cyril became patriarch of Alexandria in A.D. 412, he began a systematic program of oppression against such heretics. Because of her beliefs and her friendship with Orestes, the

* Although Suidas (ca. tenth century) implies that Hypatia was married to Isidorus of Gaza, the Neoplatonist, most historians discount this as fiction rather than fact. The romantic aspect of her life has inspired a great deal of speculation; see J. Toland, *Hypatia, or the History of a Most Beautiful, Most Vertuous, Most Learned . . . Lady* (London, 1720).

prefect of Egypt, whose influence represented the only countervailing force against Cyril, Hypatia was caught as a pawn in the political reprisals between the two factions.

Cyril was an effective inquisitor. He began by inflaming the passions of the populace, setting mobs on his detractors, leveling the synagogues, and almost completely usurping the state and authority of a civil magistrate. The turbulent mood of his own faithful and the political events that followed his actions convinced him in A.D. 415 that his own interests would be best served by the sacrifice of a virgin. At his direction, a mob of religious fanatics set upon Hypatia, dragging her from her chariot while she was on her way to classes at the university, pulling out all of her hair, and subsequently torturing her to death.

Edward Gibbon wrote (1960, p. 601)

In the bloom of beauty, and in the maturity of wisdom, the modest maid had refused her lovers and instructed her disciples; the persons most illustrious for their rank or merit were impatient to visit the female philosopher; and Cyril beheld with a jealous eye the gorgeous train of horses and slaves who crowded the door of her academy. A rumour was spread among the Christians that the daughter of Theon was the only obstacle to the reconciliation of the prefect and the archbishop; and that obstacle was speedily removed. On a fatal day, in the holy season of Lent, Hypatia was torn from her chariot, stripped naked, dragged to the church, and inhumanly butchered by the hands of Peter the reader and a troop of savage and merciless fanatics; her flesh was scraped from her bones with sharp oyster-shells, and her quivering limbs were delivered to the flames. The just progress of inquiry and punishment was stopped by seasonable gifts; but the murder of Hypatia has imprinted an indelible stain on the character and religion of Cyril of Alexandria.

Orestes felt a responsibility for Hypatia's cruel death and did what he could to bring the culprits to justice. He reported her death to Rome and asked for an investigation. Then fearing for his own life, he quit the city. The investigation was repeatedly postponed for "lack of witnesses," and finally it was given out by the Bishop that Hypatia was in Athens and there had been no tragedy. Orestes' successor was forced to cooperate with the Bishop, and as one historian phrased it, "Dogmatism as a police system was supreme" (Hubbard 1908, p. 102).

Hypatia's place in history seems relatively secure. Indeed, very often she is the only woman mentioned in mathematical histories. Her life and times have been romanticized by Charles Kingsley in his book *Hypatia: or New Foes in Old Faces* (1853), but his novel almost totally ignores Hypatia's significant work in mathematics. Neither is it to be recommended as a reliably authentic source of information, either about Hypatia or life in Alexandria during the fifth century A.D.

Mozans (1913, p. 141) on the other hand, gives more emphasis to Hypatia's place in science; he writes that she was

Among the women of antiquity what Sappho was in poetry and what Aspasia was in philosophy and eloquence—the chiefest glory of her sex. In profundity of knowledge and variety of attainments she had few peers among her contemporaries and she is entitled to a conspicuous place among such luminaries of science as Ptolemy, Euclid, Apollonius, Diophantus and Hipparchus.

He goes on to regret that this "favored daughter of the Muses" is absent from Raphael's painting *School of Athens* and attributes this omission to the fact that her achievements were not as well known in Raphael's day as they are presently. Whether Raphael's ignorance, his close relationship with the Church, or his own provincialism caused the slight, it was a phenomenon similar to that which many other women in mathematics were to experience.

For many long centuries after Hypatia's death, the science of mathematics remained relatively quiescent. Indeed, the period between the fall of Rome in A.D. 476 and the taking of Constantinople by the Turks in 1453 saw a general decline in learning and civilization.

During these long centuries, some small centers of culture developed first in one part of Europe, then in another. Italy was one such center; Gaul developed as another; and Britain, Ireland, and Germany also emerged as intellectual centers.

From the Dark Ages to the Renaissance:
The "Witch" of Agnesi
1718–1799

Over this time, when the ancient world was giving way to the medieval, a monstrous tide of misogyny had engulfed Christendom in Europe and did not commence to subside until the Renaissance began. Even in the most enlightened centers there was strong opposition to any form of higher education for females. Most would have denied women even the fundamental elements of education, such as reading and writing, claiming that these were a source of temptation and sin. (To such critics, Hroswitha, the famous nun of Gandersheim, once replied that it was not knowledge itself that was dangerous, but the poor use of it: "Nec scientia scibilis Deum offendit, sed injustitia scientis" [Mozans 1913, p. 45]).

For the most part, learning was confined to monasteries and nunneries, and these precincts guarded well the sacred mysteries of mathematics, enfranchising only those who subscribed to the religious faith of the ecclesiastics. Such schools generally constituted the only opportunity for education open to girls during the Middle Ages, and in a few of these, women were able to distinguish themselves as scholars.

Perhaps one of the most learned of these was Hroswitha, the well-known nun of the Benedictine Abbey in Saxony during the tenth century. Although she is most often cited for her dramatic compositions and as a writer of history and legend (among the latter, *The Lapse and Conversion of Theophilus* was a precursor of the famous legend of Faust), Hroswitha's writings are also an important index to the monastic mathematics of this period, and they reveal a sound intelligence of either Greek or Boethian arithmetic. In her *Sapientia*, for instance, when the emperor Hadrian inquires to know the ages of Sapientia's three daughters (Faith, Hope, and Charity), the reply is that Charity's age is represented by a defective evenly even number; Hope's by a defective evenly odd number; and that of Faith by an oddly even redundant one. It is also worthy of remark that in her writings, Hroswitha mentions four perfect numbers: 6, 28, 496, and 8128.*

Hroswitha had both the courage and originality of genius, though not all of her talents were focused on the study and development of mathematics. She was interested in various branches of learning, and her writings were an attempt to provide educational material for the women in her medieval nunnery, a purpose that motivated other scholarly nuns, including Saint Hildegard, Abbess of Bingen on the Rhine. Her capabilities in mathematics and her treatises on science earned recognition, and it has been claimed by some writers that she anticipated Newton by centuries when she wrote that the sun was the center of the firmament and its gravitational pull "holds in place the stars around it, much as the earth attracts the creatures which inhabit it" (Mozans 1913, p. 169).

* A perfect number is defined as one that equals the sum of its aliquot parts, that is, the sum of all its factors and unity: 6 = 1 + 2 + 3.

After the fall of Constantinople, there was a great influx of scholars from the famous old city on the Bosphorus into the Italian peninsula. These scholars brought with them some of the treasures of science and literature that were to help spark the interesting phenomenon we call the Renaissance. Another notable development significant to learning was the invention of the printing press with movable type. This device helped in the dissemination of knowledge and made the printed page more available to those who were not formally educated.

In Italy some traditions of the relatively free Roman matron had remained alive, but elsewhere on the European continent the status of women changed very slowly, even after the Renaissance. There were occasionally women whose talents and genius were remarkable, but the lives of these women emphasized by contrast the prevalent ignorance of the great mass of women who had little access to instruction even in its most fundamental form.

France and Germany saw a revival of the antifeminist crusade that had stifled women's aspirations in ancient Greece and Rome. The Teutonic mentality did not recognize intelligence in women; Luther was a strong influence in his opposition to the education of females.

In England Henry VIII had destroyed the conventual system, leaving women without any systematic education for a long period. Elizabeth I did nothing for the education of females; where their intellectual progress was enhanced at all during these years, it was due to private tutoring or the protracted efforts of individual women for their right to knowledge. Here, as in most of Europe, women were in many respects even further removed from knowledge than they were during the Dark Ages.

But on the Italian peninsula, where the Renaissance had its origin, some Italian women had made their mark on the academic world, even before the close of the Middle Ages. Some had earned doctorates and had become lecturers and professors in the universities of Bologna and Pavia.

The advent of the Renaissance signaled the return of many Italian women to an active role in the educational movement. One historian (Mozans 1913, p. 58) wrote of this period in Italian history,

The universities which had been opened to them at the close of the Middle Ages, gladly conferred upon them the doctorate, and eagerly welcomed them to the chairs of some of their most important faculties. The Renaissance was, indeed, the heyday of the intellectual woman throughout the Italian peninsula —a time when women enjoyed the same scholastic freedom as men.

Nor were these women scholars exposed to ridicule. This same historian wrote (Mozans 1913, p. 63) that the men of those days

. . . were liberal and broad-minded . . . who never for a moment imagined that a woman was out of her sphere or unsexed because she wore a doctor's cap or occupied a university chair. And far from stigma-tizing her as a singular or strong-minded woman, they recognized her as one who had but enhanced the graces and virtues of her sex by the added attractions of a cultivated mind and a developed intellect. Not only did she escape the shafts of satire and ridicule, which are so frequently aimed at the educated woman of today, but she was called into the councils of temporal and spiritual rulers as well.

Woe betide the ill-advised misogynist who should venture to declaim against the inferiority of the female sex, or to protest against the honors which an appreciative and a chivalrous age bestowed upon it with so lavish a hand. The women of Italy, unlike those of other nations, knew how to defend them-selves, and were not afraid to take, when occasion demanded, the pen in self-defense. This is evidenced by numerous works which were written in response to certain narrow-minded pamphleteers or pitiful ped-ants who would have the activities of women limited to the nursery or the kitchen.

The talent and genius that flowered as a result of this enlightened attitude was enormous: Women became famous in the arts, in medicine, literature, philosophy, science, and languages, and there were also important names surfacing in mathematics dur-ing the seventeenth and eighteenth century. There were Tarquina Molza, who was taught by the ablest scholars and was honored by the senate of Rome for her accomplishments; Maria Angela Ardinghelli of Naples; Clelia Borromeo of Genoa, who was widely praised and of whom it was said that "no problem in mathematics and mechanics seemed to be beyond her comprehension" (Mozans 1913, p. 142). There were Elena Cornaro Piscopia, honored by the University of Padua for her proficiency in mathematics; Laura Bassi, primarily known for her work in physics (her work centered on Descartes and Newton, and she was a member of the Bologna Academy of Sciences); and Diamente Medaglia, who wrote a special dissertation on the importance of mathematics in the curriculum of studies for women and is quoted: "To mathematics, to mathematics, let women devote attention for mental discipline" (Mozans 1913, p. 142).

Some of these women were mathematicians in the most rigorous sense; others worked on the periphery of the discipline, but whatever their proper designation, each had absorbed enough mathematics to make her efforts an exemplary part of that enlightened and powerful age we call the Renaissance, and for this these women have earned a separate recognition, a separate warrant of our attention.

Far more remarkable than any of these women, however, was Maria Gaetana Agnesi, called one of the most extraordinary women scholars of all time. She was born in Milan on May 16, 1718, to a wealthy and literate family; like Hypatia's, her father was a professor of mathematics. Dom Pietro Agnesi Mariami occupied a chair at the University of Bologna, and he, along with Maria's mother, Anna Brivia, very carefully planned the young girl's education so that it was rich and profound.

She was recognized as a child prodigy very early; spoke French by the age of five; and had mastered Latin, Greek, Hebrew, and several modern languages by the age of nine. At around this age, she delivered a discourse in Latin defending higher education for women, a subject that continued to interest her throughout her life.

Maria's teen-age years were spent in private study and in tutoring her younger brothers (she was the oldest of 21 children). During this time, she also mastered the study of mathematics as it had been developed by such masters as Newton, Leibniz, Fermat, Descartes, Euler, and the Bernoulli brothers.

The Agnesi home was a watering place for a select circle of the most distinguished intellectuals of the day, and Maria acted as hostess for her father's carefully chosen assemblies. She participated in the seminars among those gathered in her father's study by presenting theses on the interesting philosophical questions under discussion, and her father encouraged her to engage in disputations with these scholars.

Monsieur Charles De Brosses, the president of the parliament of Burgundy, wrote in his *Lettres sur l'Italie* about one of these seminars to which he and his nephew were invited (Agnesi 1801, p. 13). He was particularly impressed with Maria's erudite versatility in the discussion of such diverse subjects as the manner in which the soul received impressions from corporeal objects, and in which those impressions are communicated from the eyes and ears and other parts of the body on which they were first made, to the organs of the brain which is the general sensorium or place in which the soul receives them; we afterwards disputed on the propagation of light and the prismatic colours. Loppin then discoursed with her on transparent bodies, and curvilinear figures in Geometry, of which last subject I did not understand a word. . . . She spoke wonderfully well on all of these subjects though she could not have been prepared before-hand to speak on them, anymore than we were. She is much attached to the Philosophy of Sir Isaac Newton; and it is marvelous to see a person of her age so conversant with the abstruse subjects. Yet, howevermuch I may have been surprised at the extent and depth of her knowledge, I have been much more amazed to hear her speak Latin (a language which she certainly could not often have occasion to make use of) with such purity, ease and accuracy.

De Brosses mentions that this particular party was attended by about thirty people from several different nations of Europe, seated in a circle, questioning Maria. Pietro Agnesi was understandably proud of his accomplished daughter, but these displays were contrary to her shy, bashful nature, and she prevailed upon her father to give them up when she was around twenty years old. At about this time she began to express a desire to enter a convent so that she might spend her life in sequestered study and work with the poor. This request was denied by her father.

Maria never married. She gave most of her time to the study of mathematics, to caring for her younger brothers and sisters, and (after her mother's death) to assuming the duties of the household.

In 1738 she published a collection of complex essays on natural science and philosophy called *Propositiones philosophicae*, based on the discussions of the savants who had gathered in her father's home. Again, these essays expressed her conviction that women should be educated in a variety of subjects.

By the age of twenty, she had entered on her most important work, *Analytical Institutions*, a treatise in two large quarto volumes on the differential and integral calculus. She spent ten years on this work, and her natural talent for mathematics may be reflected in her report that on several occasions during this time, after working all day on a difficult problem that she could not solve, she would arise at night and while in a somnambulistic state, write out the correct solution to the problem.

When her work was finally published in 1748, it caused a sensation in the academic world. Although she had originally begun the project for her own amusement, it had grown, first into a textbook for her younger brothers and then into a more serious effort. It has the distinction of being one of the most important mathematical publications produced by a woman up until that time. It was a classic of its kind and the first comprehensive textbook on the calculus since l'Hôpital's early book.* It was also one of the first and most complete works on finite and infinitesimal analysis and was not superseded until Euler produced his great texts on the calculus later in the century.

Agnesi's great service was that she pulled together into her two volumes the works of various mathematicians, including Newton's method of "fluxions" and Leibniz's method of differentials. These and other works concerning analysis were scattered through the writings of various authors, some printed in foreign journals. Maria's scholarship and her facility with languages helped her to collect these into a compendium that saved students the complicated task of seeking out developments and methods formerly dispersed in a variety of sources. Her volumes were translated into French and English and were widely used as textbooks.

The first section of *Analytical Institutions* deals with the analysis of finite quantities and discusses the construction of loci, including conic sections. It also deals with elementary problems of maxima and minima, tangents, and inflections.

The second section is devoted to the analysis of

* Although the Marquis de l'Hôpital is given full credit for his *Analyse des infiniment petits*, the wife of this illustrious mathematician was of special assistance in preparing this work for the press. She also shared her husband's genius for mathematics and worked closely with him; indeed, she saw that most of his important work was published after his death.

"infinitely small quantities," quantities defined as so small that when compared to the independent variable, the proportion is less than that of any assigned quantity. (If such infinitesimals, called "differences" or "fluxions," are added to or subtracted from the variable, the difference would not be significant. "Differences" or variables tending to zero, and "fluxions," or finite rates of change, are treated here as essentially the same quantities.)

The third section of Agnesi's work deals with the integral calculus and gives a general idea of the state of knowledge concerning it at the time. She gives some specific rules for integration, and there is a discussion on the expression of a function as a power series. The extent of convergence is not treated.

The last section of the volume discusses "inverse method of tangents" and very fundamental differential equations.

But many of the other important aspects of *Analytical Institutions* have been eclipsed by Agnesi's discussion of a versed sine curve, originally studied by Fermat. This plane cubic curve has the Cartesian equation $xy^2 = a^2(a - x)$. (Agnesi begins with the geometrical principle that if the abscissa of corresponding points on a curve is equal to that of a given semicircle, then the square of the abscissa is to the square of the radius of the semicircle in the same ratio as that in which the abscissa would divide the diameter of the semicircle.) This curve had been studied earlier by Guido Grandi, as well as Fermat. It had come to be called a *versiera*, a word derived from the Latin *vertere*, "to turn," but it was also an abbreviation for the Italian word *avversiera*, or "wife of the devil."

In 1801, when Maria's text was translated into English by John Colson, professor of mathematics at Cambridge, Colson rendered the word *versiera* as witch, and through this or some such mistranslation, the curve discussed by Maria came to be known as the "witch of Agnesi." Subsequently, where mention of this woman is made in modern English textbooks, it is most often by this phrase. The exquisite irony of this term is not lost on those who are familiar with Agnesi's life of selfless service and piety.

Maria's books attracted the attention of the French Academy of Sciences, and a committee was appointed to assess them. A deputy wrote her afterward (Beard 1931, p. 442),

I do not know of any work of this kind that is clearer, more methodical or more comprehensive. . . . There is none in mathematical sciences. I admire particularly the art with which you bring under uniform methods the divers conclusions scattered among the works of geometers and reached by methods entirely different.

Despite this tribute, however, the French Academy did not admit Agnesi. Its constitution barred females, despite the fact that the very notion of the Academy was introduced to its founder, Richelieu, in the salon of a woman, Madame de Rambouillet.

Fortunately, Italian academics were more liberal, and Maria was elected to the Bologna Academy of Sciences. There were also many other honors: Her book had been dedicated to the Empress Maria Theresa, who showed her appreciation by sending Maria a splendid diamond ring and a small crystal casket set with diamonds and precious stones.

But the recognition that pleased her most came from Pope Benedict XIV. He was interested in mathematics, and he recognized the exceptional ability of Maria Agnesi. His letters indicate his respect for her accomplishments, and it was through his invitation that she was given an appointment as honorary lecturer in mathematics at the University of Bologna.

Her name was added to the faculty roll by the senate of the university, and a diploma to this effect was sent her by the pontiff. The diploma was dated 5 October 1750. Sister Mary Thomas a Kempis wrote later that Agnesi's name remained on the university's "Rotuli" until 1795–1796.

There is some difference of opinion among historians as to whether Agnesi accepted this appointment or not. She was urged to do so by many of her contemporaries, including the famous physicist Laura Bassi. Most reviews of her life indicate that she occupied the Chair of Mathematics and Natural Philosophy at Bologna from 1750 to 1752; other writers say she only filled in for her father during his last illness; still others insist that she eschewed the pontiff's offer, preferring instead to remain in her beloved Milan. In retrospect, it would appear that she did accept this position and served at the university until her father's death in 1752, when she decided to return to a quieter life of study and comparative solitude.

She relinquished the ambition to do any further work in mathematics; when, in 1762, the University of Turin asked her for her opinion of the young Lagrange's recent articles on the calculus of variations, her response was that she was no longer concerned with such interests.

True to her deeply religious nature, she began to devote most of her time to charitable projects with the sick at the hospital of Maggiore and with the poor of her parish, San Nazaro.

Sister Mary Thomas a Kempis, whose beautiful article *"The Walking Polyglot"* reviews these charitable efforts of Agnesi's, reports that, "To extend her work more and more she saved on her dresses, on her meals, and on her dear books, she did not hesitate to sell her imperial gifts and even the crown set with precious jewels given her by Pope Benedict XIV" (Thomas a Kempis, p. 216).

She turned her home into a refuge for the helpless and the sick, the aged, and the poor. Neglected women were cared for in her own rooms when there were no other facilities. And when the Pio Instituto Trivulzio, a home for the ill and infirm, was opened in 1771, the archbishop asked Maria to take charge of visiting and directing women, particularly the ill. She took on this duty in addition to the burden of maintaining her own small hospital. When these duties became too burdensome, she took up full-time residence at the Institute in 1783, insisting upon paying rent so as not to diminish the capital of the poor. The annals of the Institute call her "an angel of consolation to the sick and dying women until her death at the age of eighty-one years on January 9, 1799" (Thomas a Kempis, p. 216).

Agnesi was buried in a cemetery outside the Roman gate of the city walls. She shares a common grave with fifteen old people of the Luogo Pio. There is no elaborate monument over her tomb, nor is one needed. She has been widely honored (and continues to be) for her good works.

On the one-hundredth anniversary of her death, Milan took note of her life: streets in Milan, in Monza, and in Masciago were given her name. A cornerstone has been placed in the facade of the Luogo Pio, the inscription on it proclaiming her "erudite in Mathematics glory of Italy and of her century." A normal school in Milan bears her name, and scholarships for poor girls have been donated in her honor.

Today, almost two hundred years after her lifetime of hard work, her memory is still vital and inspiring.

France, during the post-Renaissance period, offered little opportunity for the education of females; although there were a few learned Frenchwomen during the seventeenth century, their number was quite small. With few exceptions, Frenchwomen showed little inclination to spend long hours at study, and still fewer shared with their sisters south of the Alps the disposition to attempt a scholarly career.

However, the traditions of the Renaissance were still quite strong among certain classes of the aristocracy, and during the reign of Louis XIV, the Institut de

Emilie de Breteuil,
Marquise du Châtelet
1706–1749

Saint-Cyr, the first state school for girls, was founded to educate the daughters of the nobility. Mme de Maintenon, the king's morganatic wife, was responsible for starting the Institut, and although she was one of the most enlightened women of her time, the school offered studies of only a very elementary character. The purpose of St. Cyr's was to prepare future wives for the nobility; the physical and natural sciences were, of course, considered to be outside the interests, needs, and capabilities of such women. During the seventeenth century, J. Molière introduced his two devastating plays *Les Précieuses ridicules* and *Les Femmes savantes*; and N. Boileau also published his *Contre les Femmes*. All of these works ridiculed learned women and had made the term "femme savante" a damaging epithet.

On Molière's part, it was claimed that his wickedly effective satire had been aimed at superficial pedantry and not at the genuinely educated woman. But because his wit and humor had reinforced the prevailing mythology about women, its effect was a more diffused and general one; it helped to create a social plenum quite hostile to the display of any knowledge on the part of women. So severe was this general hostility that even women of real learning took great pains to hide their intellectual activity. If a woman dared to devote herself to science or philosophy, the salon offered the only haven for discussing such topics; but even there the talk had to be kept gay, bright, and shallow. Women were considered to be incapable of abstraction, generalization, or the mental concentration necessary to comprehend such subjects as mathematics and the physical sciences, and they were therefore excluded from these studies. Jacques Rousseau, his sensibilities blunted by tradition, thought that women should relate exclusively to practical and domestic matters, that abstract and speculative truths and principles were beyond their capabilities. He is quoted as believing "everything that tends to generalize ideas is outside of her competence" (Mozans 1913, p. 92). He was joined in this sentiment by such popular figures as D. Diderot, Baron Montesquieu, Voltaire, and the Encyclopedists, as well as Molière and Boileau.

These men disliked the notion of education for the daughters of the nobility, and education for the daughters of the poor was beyond any consideration at all. It is ironic that most of these men sought out sophisticated and intellectually stimulating women as their own companions; in some cases, they were openly indebted to these women for the intellectual assistance they gave. This is true most specifically in the case of Voltaire.

Few Frenchwomen were redoubtable enough to fly in the face of such sentiment, and few had the courage to defend those who tried. For the next two centuries, France produced scarcely any women whose claim to fame was based on anything more substantial than the social life of the salons, which reputedly sparkled with intellectual fire though not of a very scholarly nature.

This, then, is the milieu into which Emilie de Breteuil was born in Paris on December 17, 1706. Her father was Louis Nicholas le Tonnelier, baron de Breteuil, a fashionable if somewhat shallow figure who was Introducteur des Ambassadeurs, or head of the protocol at Court.

Emilie's mother was convent educated, and her manners were those of the great ladies of the day. Her instructions to her children were limited. "Do not blow your nose on your napkin. . . . Break your bread and do not cut it. Always smash an egg-shell when you have eaten the egg. . . . Never comb your hair in church" (Mitford 1957, p. 17). It is doubtful that Emilie listened to such drivel, pragmatic though it was in her circle. However she did learn well (perhaps from her beautiful mother) the exacting, systematic art of coquetry, which was so popular at the time. There is also evidence that she developed other personal charms of a novel order.

She began very early to show enough promise to convince her father that she was a genius who needed some attention, and she was provided with a relatively good education by the existing standards. She was a natural linguist and quickly mastered Latin, Italian, and English. She studied Virgil, Tasso, and Milton, and she translated the *Aeneid.* She could recite long passages from Horace and studied the works of Cicero.

But her real, true, and lasting love was mathematics; she was encouraged in this study by a family friend, M. de Mezieres, who recognized her genius. Voltaire wrote later of her facility with numbers, "I was present one day when she divided nine figures by nine other figures, entirely in her head, without aid of any sort, and an astonished geometrician was there who could not follow what she did" (Hamel 1910, p. 70).

Emilie's precocity also showed itself in other ways. She had what writers were wont to call in the lexicon of the times a somewhat "passionate nature"; it is rumored that she never lacked for romantic attachments either before or after her marriage at nineteen to the thirty-four-year-old Marquis du Châtelet. Many of her biographers have dealt at length with such affairs, and it is significant that in reviewing Emilie's life, they have been seduced by the temptation to overemphasize the trivial, the anecdotal, and the chimerical at the expense of the more scholarly aspects of her career. It is true her work in mathematics was seldom as original or as innovative as that of many other women who are the subjects of this book, but it was substantive, and the fact that it was accomplished at all is worthy of considerable note.

Emilie's husband was the head of an old Lorraine family. He was the colonel of a regiment and was often away on garrison duty. During the first two years of their marriage, Emilie gave birth to two children, a boy and a girl, and then when she was twenty-seven, another boy was born to her. Neither the children nor the Marquis du Châtelet deterred her from grasping life with a headlong passion that was unusual, even for those days, in the French court. Here, Emilie enjoyed privileges usually reserved for duchesses: of sitting in the presence of the queen, and of traveling in her suite. She loved the social life of the court, especially the gambling and the amorous adventures.

Emilie was never as circumspect or as discreet as she should have been to avoid gossip; it testifies to the strength of her spirit that she was able to live almost as a law unto herself despite the criticism of others. She was honest and honorable according to her own ethical code (though that code quite often violated the standards and customs in vogue at the time).

She was not altogether unconcerned about the opinions of others, but she was determined to live according to her own needs and, as far as possible, without artificial constraints. It is to her credit that she never struck back at convention, despite the gossip that often trailed her.

Emilie's nature was a complex one, reckless, enthusiastic, changeable, positive, and direct. She lived life at full tilt like a spirited, healthy child. One biographer wrote of her, "Not one of the frivolous joys of life was too frivolous for her. The activity of her mind and the natural simplicity of her character occasioned a bizarre struggle between work and play" (Hamel 1910, p. 131). She committed two unforgivable sins that aroused both jealousy and resentment against her. First, she refused to give up her serious study of mathematics and engaged the most advanced tutors available to help her in this endeavor. Second, she stole the heart of Voltaire, the most facile wit who ever illumined a salon, and she appropriated him as her constant companion for the rest of her life. These two pursuits were sufficient to arouse considerable petulance in any prominent salon ennuyée.

Emilie's husband cared little about her intellectual proclivities and was no doubt relieved that her liaison with Voltaire furnished her an opportunity to indulge these interests. She had had affairs with other men, including the Marquis de Quebraint and the Duc de Richelieu, a notorious womanizer. Indeed, it was to the latter that Emilie appealed for help in convincing her husband that there should be no fuss about her affair with Voltaire.

During this affair, Voltaire's *Lettres sur les Anglais* were published without his authorization; these so inflamed certain French sensibilities that there were rumors his banishment was imminent. Concerned for his safety, Emilie suggested that they leave Paris for Cirey, the ancestral home of the du Châtelet family, near the Lorraine border.

It was here at Cirey-sur-Blaise that the two lived,

working intermittently, away from the turbulence of Paris and court life. The castle at Cirey was old and neglected, but as Voltaire (Hamel 1910, p. 57) described their idyll there,

This old chateau she ornamented and embellished with tolerably pretty gardens; I built a gallery and formed a very good collection of natural history; in addition to which we had a library not badly furnished.

We were visited by several of the learned, who came to philosophize in our retreat; among others we had the celebrated Koenig. . . . Maupertuis came also, with Jean Bernoulli. . . . In this our delightful retreat we sought only instruction, and troubled not ourselves concerning what passed in the rest of the world. We long employed all our attention and powers upon Leibniz and Newton; Mme du Châtelet attached herself first to Leibniz, and explained one part of his system in a book exceedingly well written, entitled *Institutions de physique*. She did not seek to decorate philosophy with ornaments to which philosophy is a stranger; such affectation never was part of her character, which was masculine and just. The qualities of her style were clearness, precision, and elegance. If it be ever possible to give the semblance of truth to the ideas of Leibniz, it will be found in that book; but at present few people trouble themselves to know how or what Leibniz thought.

Born with a love of truth, she soon abandoned system, and applied herself to the discoveries of the great Newton; she translated his whole book on the principles of the mathematics into French; and when she had afterwards enlarged her knowledge, she added to this book, which so few people understand, an "Algebraical Commentary," which likewise is not to be understood by the general reader.

It is rare that two extravagantly gifted and exacting people have been able to perpetuate an alliance of interdependence and cooperation such as Emilie and Voltaire enjoyed from 1733 to 1749. Very few great men or women have been fortunate enough to find so perfect a companion or one who possessed so many qualities in common.

Voltaire and Emilie were each a rare amalgam of diverse and contradictory traits. They were essentially intellectual, yet piquant; they were driven and consumed by scholarly curiosity, yet they were urbane, ardent, and convivial; both were mercurial, yet they were generous, charitable, and ungrudging.

Two such bifurcated and hearty personalities, both impassioned to define truth in their own particular idiom, were destined to have difficulty accommodating to each other; but these difficulties were not of a major order (even when magnified by gossips), and their relationship was a sound and mutually profitable one for long years of work and leisure together.

To assess Emilie's accomplishments properly, it is necessary to understand the social context within which she lived and worked. One of Emilie's most influential tutors was Pierre Louis de Maupertuis, one of the leading mathematicians and astronomers of the day. He was a fellow of the English Royal Society and a member of the French Academy of

Sciences and served for a time as president of the Berlin Academy.

Both Maupertuis and Voltaire were attempting to move French thought beyond Descartes. Toward this end, Voltaire's *Lettres* had been written, and Maupertuis had also published several papers in an effort to convert French academic thought away from Descartes and toward the ideas of Newton. Both men appreciated the fact that Descartes had done much to put scientific work on a more rigorous mathematical basis, but by the beginning of the eighteenth century, Cartesian ideas were being looked on by some as representing a relatively timid philosophy.

In England, Newton's *Principia* had been published in 1687. It had had very little immediate effect, for it could be read only with difficulty even by mathematicians; it was intelligible to only a very few of Newton's contemporaries who were not much interested in promoting it to less scientific readers. However, by the time Voltaire and Maupertuis visited England in the early part of the 1700s, both were impressed with the scientific advances made by the English and credited these advances to Newtonism. When they returned to France, these enthusiastic disciples set about trying to get Newton's ideas accepted in French academic circles, which were largely under the influence of a fierce nationalism at the time.

It is doubtful that Voltaire actually had a very profound understanding of Newtonian principles;

Maupertuis had perhaps a better grasp, judging from his several papers on the subject. But it is interesting and ironic to note the different impacts their efforts produced. Maupertuis, the more scholarly mathematician, succeeded in making Newtonian ideas a rather modish topic of conversation in the great salons of France, while Voltaire was to influence his friend Emilie du Châtelet to translate Newton into French, making his work available to French mathematicians and scholars. Although not published until after her death, Emilie's translation and its accompanying notes did much to free French thought from its allegiance to Cartesianism.

At Cirey, Emilie and Voltaire had established a well-equipped laboratory in which they took special delight. One of Emilie's first scientific works was concerned with an investigation regarding the nature of fire. In 1738, the French Academy of Sciences held a competition for the best essay on the subject; although the announcement had been made a full year ahead, Emilie did not elect to enter until a month before the deadline of September 1. Voltaire had been working on his essay long before this time, but the impetuous Emilie entered only after she began to disagree with the points Voltaire was attempting to make. She kept her entry secret from Voltaire and completed it at night, sleeping an hour or so each day, and keeping herself awake by plunging her hands into iced water.

Mme de Graffigny, a visitor at Cirey at the time,

read both essays and wrote of them, "It is true that when women mix themselves up with writing they surpass men. What a prodigious difference! But how many centuries does it take to produce a woman like her?" (Hamel 1910, p. 161).

Both essays were good and both presented cogent and original ideas. Emilie's work anticipated the results of subsequent research by arguing that both light and heat have the same cause or are both modes of motion; she also discovered that different-colored rays do not give out an equal degree of heat.

Neither of the essays won the competition, however, for the prize was divided among three other entries, part of it going to Euler. At least Emilie and Voltaire had the comfort of knowing that they had failed in good company, and there were consolations. Indeed, the Academy was so impressed with the originality of their entries that they were printed at the close of the Prize Essays and Prince Frederick wrote quite flattering letters of praise for their contributions.

Emilie was a resolute, almost unrelenting scholar, but she made impossible demands on her tutors. Her swift mind outpaced them; her irregular hours disrupted their lives; her rigorous questions were frequently impossible to answer. In 1739, she quarreled powerfully with Samuel Koenig, a protégé of Maupertuis, who had been engaged to tutor her. Their dispute was a philosophical one over the subject of the infinitely small, and since no agreement could be reached, the two decided to end their association.

Koenig was to have his revenge, however, for when, in 1740, Emilie's book *Institutions de physique* was published, he told everyone in Paris that this was only a rehash of his own lessons to Mme du Châtelet. Emilie was furious, and because she had discussed her ideas with Maupertuis long before Koenig had been engaged as her tutor, she appealed to him and to the Academy of Sciences to help her prove the book was indeed her own.

Most of the knowledgeable scientists agreed that she was more than capable of the work, but she was left with the feeling that she had not received the support she deserved, that being a woman had worked against her, and that Koenig had thrown some doubt on her honesty. She was somewhat vindicated when, in 1752, long after her death, Koenig showed his true colors: he published a forged letter from Leibniz and became embroiled with the Berlin Academy. Maupertuis also was to quarrel with Koenig over his perfidy, but again this came too late to help Emilie.

According to her own account, Emilie had intended to produce *Institutions* as an essay on physics for her son. The classic physics textbook in use at the time was Rohault's, which had been written over eighty years before, and Emilie saw a need for a modernized, supplementary text dealing with the problems discussed by the seventeenth-century physicists, the methods they had used, the results they had achieved, and the problems they had posed. All of this entailed far more than could be covered in an essay, however, and she produced instead a comprehensive textbook within a framework not unlike that of a modern text with introductions, definitions, historical development of concepts, and methods of thinking about physical phenomena. In addition, *Institutions* also dealt with a set of metaphysical principles, including the five principles of reasoning, the nature of matter, the nature of knowledge, and so forth.

It is sometimes said of this work that Emilie attempted to synthesize the metaphysical aspects of Leibniz with the physical aspects or ideas of Newton. It is true that *Institutions* did describe with clarity and detail the role of Leibniz and Newton (as well as Descartes) in the establishment of the modern physics of the times, but the science Emilie dealt with encompassed the efforts of human thought among other scientists, French, English, German, Swiss, and Dutch, as well. She introduced physics as a general subject and attempted to define particular points concerning space, time, extension, and so forth. She gave the historical backdrop as constituted by the work of Descartes, C. Huyghens, Kepler, and even the Greek atomists and introduced new material by more recent scientists, such as Bradley, Clarke, Wolff, and others.

She was not provincial in her treatment and did not accept indiscriminately the position of any one scientist. Rather she took the position that what was needed was a better understanding of the achievements of each, as well as a method of thinking (or using hypotheses) scientifically. She pointed out that Newton had been overzealous in condemning hypotheses, while on the other hand, Descartes had excessively stressed the doctrine of intuition. She brought into focus the fact that great thinkers sometimes err in their extremes and abhorred the fact that the doctrines of Descartes and Newton had become polar rallying points for seventeenth-century science.

Her work was not doctrinaire and cannot honestly be judged to have been a polemic in defense of Leibniz, Newton, or Descartes. Voltaire seems to have had some investment in perpetuating the fiction that she was biased toward Leibniz in the beginning and only later came to Newtonism. A close reading of *Institutions* reveals that she took a much more scholarly and methodical approach in her effort to trace the growth of modern physics and to summarize the thinking of the philosopher-scientist of her century.

Institutions was written in a vigorous, almost bleak style. The work established Emilie's competence among such contemporaries as Clairaut, the Bernoullis, Mairan, and Maupertuis, despite Koenig's efforts to claim part of the credit for her efforts.

The years Emilie shared with Voltaire at Cirey marked the most productive period in her life. Voltaire's political activity and his popularity, his widely quoted witticisms, and his light-hearted antics have often obscured the feverish intellectual activity that went on during the sojourn of the "two philosophers" at Cirey. But this activity was intense, particularly on Emilie's part, despite frequent interruptions by visitors from the capitols of Europe and by the vagaries of Voltaire's political life. There was a diversity in these mental exertions that would not have been possible had Emilie and Voltaire not been working in unison. The intellectual mix was mutually established and mutually profitable, and although the period was a relatively more productive one for Emilie than for Voltaire, he too benefited from this association. Ira O. Wade (1969, p. 276) has written of this period in Voltaire's career that "Mme du Châtelet's contribution to the development of Voltaire's thought . . . was substantial." And further, that "*Candide* would have had less chance to be the *Candide* it is had Mme du Châtelet not existed." Wade goes on to observe that "At all events, the importance of the *Institutions de physique* and the translation of the *Principia* upon the intellectual development of Voltaire cannot be exaggerated."

Emilie's husband had become fond of Voltaire, and he was most cooperative in adjusting to the ménage à trois. However, he disliked the hours that the two kept when they were working, and he often preferred to take his meals with his son. Although the Marquis du Châtelet seems to have been quite proud of his wife and her accomplishments, it was evident that he did not require her constant company.

Longchamp, a servant at Cirey who later became a writer, described life there (Hamel 1910, p. 126):

Mme du Châtelet passed the greater part of the morning with her writings, and did not like to be disturbed. When she stopped work, however, she did not seem to be the same woman. The serious air gave place to gaiety and she gave herself up with the greatest enthusiasm to the delights of society. She might have been taken for the most frivolous woman of the world. Although she was forty years old, she was always the life of the company, and amused the ladies of society who were much younger.

A visitor, the Abbé de Breteuil, wrote that when there were no guests present, Emilie and Voltaire remained tied to their desks, the former working during the greater part of the night as well as all day long. She slept only two or three hours and left off working for a cup of coffee during the day, then returned to work again. She was always spectacularly healthy; Voltaire was less so, but he too was a slave to his work.

There were interludes away from Cirey, particularly at the court of the Polish ex-king Stanislas at Luneville, where Emilie was quite capable of dancing, feasting, gambling all night, then getting up to work out mathematical problems before breakfast. She was an inveterate gambler, but she was never very lucky at cards. Voltaire opposed her gambling, and she attempted to express her feelings about this (and other) pastimes in an essay called "Reflections sur le Bonheur." She claimed that the only pleasures left to a woman when she is old are study, gambling, and greed. Gambling with high stakes "shakes up the soul and keeps it healthy" (Mitford 1957, p. 188).

It was at Luneville in the early spring of 1748 that Emilie met and fell in love with the Marquis de Saint-Lambert, a courtier and very minor poet. He was never far from her thoughts from then until the end of her life, although he obviously did not return her passion and was nowhere near her intellectual equal.

At first, Voltaire knew nothing of her infatuation, and it was only by accident that he discovered her intimacy with Saint-Lambert. Although he was piqued, it did not destroy the friendship between them, and when Emilie found that she was pregnant with Saint-Lambert's child, it was Voltaire who helped her to deceive her husband.

In one of the most dishonorable acts of her life, Emilie (along with Saint-Lambert and Voltaire, both of whom had gone back to Cirey with her for the purpose) tricked her husband into believing that the child was his own. It is doubtful that du Châtelet was completely taken in by their charade, however, for Emilie's "secret" was even at that time widely known both at Luneville and at Cirey and was the subject of a great deal of gossip among their friends.

The Marquis was, as always, affable and obliging, and Emilie's fears were quieted. She could even laugh at Voltaire's piquant little joke when he told her that since the child had no claim to a father, it should be classed among her "miscellaneous works."

In June of 1749, Emilie had finished her work with Clairaut, an old friend with whom she had been studying, but she still had not completed her book on Newton. She rose at 9:00 A.M. and worked until 10:00 P.M., when she had a light dinner with Voltaire. She gave up most of her social life and saw few of her friends. She was determined to finish this work and wrote to Saint-Lambert (Hamel 1910, p. 363): "Do not reproach me with it. I am punished enough without that. I have never made a greater sacrifice to reason. I *must* finish it, though I need a constitution of iron." At the time, she was close to her forty-third birthday and pregnant.

She wished to have her child in the palace at Luneville, and since this was agreeable to Stanislas, it was here that she spent the summer of 1749. Her daughter was born in early September of that year.

As Voltaire describes it, the little girl arrived while her mother was at her writing desk, scribbling some Newtonian theories, and the newly born baby was placed temporarily on a quarto volume of geometry, while her mother gathered together her papers and was put to bed.

For several days, Mme du Châtelet seemed to be well and happy. Voltaire, Saint-Lambert, and her husband were all in attendance, and they assembled in her room daily to keep her company while she convalesced.

But in the late afternoon of September 10, 1749, Emilie died quietly and suddenly. Voltaire, who was with her at the end, was distraught and in tears. He stumbled from the room and fell at the outside door leading to the terrace. Saint-Lambert, who had followed him, lifted him up and Voltaire said gently and sadly, "Ah! Mon ami! C'est vous qui me l'avez tuée" (Mitford 1957, p. 270).

Emilie's little girl also died a few days afterward. One of Emilie's sons had died earlier; her first daughter was married to an old Neapolitan duke; her last remaining son, who was made a duke and who served as ambassador to London from 1768 to 1770, died on the guillotine at the age of sixty-six. His son died during the Revolution; and so Emilie's family became extinct. Only her work lives on.

Of her accomplishments, Mozans (1913, p. 202) writes

All. things considered, the Marquise du Châtelet deservedly takes high rank in the history of mathematical physics. In this department of science she has had few, if any superiors among her sex. And, when we recollect that she labored while the foundations of dynamics were still being laid, we shall more readily appreciate the difficulties she had to contend with and the distinct service which her researches and writings rendered to the cause of natural philosophy among her contemporaries.

Her greatest achievements as a mathematician and in the general area of science were her *Institutions de physique* and her translation of Newton, which was published posthumously in 1759 with a "Preface historique" by Voltaire. This included a set of her mathematical analyses of Newton's third book on his "système du monde" and material on Clairaut's demonstration of Newton's theorem that all curves of the third order are projections of one of five parabolas.

Other works were her paper on fire, which she presented to the Academy of Sciences in 1738, and a manuscript on optics. Of this latter work, only the fourth section is extant, but it is possible from this portion to reproduce the general structure of the entire work. In addition to these achievements, Mme du Châtelet was also interested in critical deism, particularly where this deism touched on the science of her century. She produced a manuscript based on her critical examination of the Old and New Testaments, and although this manuscript was never published, scholars are now examining it, specifically for its significance to the development of Voltaire's thought.

Emilie's essay "Discours sur le Bonheur," called by Voltaire, "essays on happiness," was her single contribution in the sphere of moral philosophy. But in this work, it is possible to get a clearer notion of the forces, the subtleties, and the passions that drove Emilie du Châtelet in her zesty love of life, her love of love, and her love of study.

Caroline Herschel is cited in scientific annals more often for her work in astronomy than for her mathematical acumen. This is understandable considering her accomplishments in astronomy, but she rightfully deserves recognition in both fields. Although she never received any formal mathematical training, this fact only serves to emphasize the magnitude of her accomplishments and the strength of her determined spirit.

Caroline Lucretia Herschel was born at Hanover, Germany, on March 16, 1750, while this territory was

Caroline Herschel
1750–1848

still part of the British crown. Her father, Isaac Herschel, a musician in the Hanoverian guard, encouraged the development of Caroline's musical talents, and she learned to play the violin competently enough to participate in concerts. The balance of her education was neglected. Although she learned to read and write, there was little exposure to arithmetic, and her mother disapproved actively of this emphasis on learning.

Her mother was an industrious hausfrau, frugal, hard-working, undemonstrative; Caroline grew up with scarcely any expressions of affection from her. However, both parents were influential in developing uncommonly strong qualities of character in Caroline and in her brother, William, who was eleven years her senior.

Isaac Herschel warned his daughter relentlessly that her lack of physical beauty and fortune would prevent her from finding a suitable husband, at least until she grew much older and her true character was reflected in her face. Caroline, who docilely accepted her father's evaluation, had no expectations along these lines and, indeed, never married.

After her father's death in 1767, Caroline was faced with the problem of making her own way in the world, and she began to consider her options. She could do housework, and she had some musical training, but neither of these were sufficient to warrant employment. If she could only learn to embroider, her qualifications as a governess might be enhanced, and she persuaded an invalid neighbor girl to teach her this art. However, even this activity had to be kept from her mother, who demanded arduous household duties of her, so the lessons took place early in the morning before the day's work began. Caroline's requests to attend school were dismissed as impractical and unnecessary, although she did eventually win her mother's permission to sew.

She was finally rescued when her brother, William, who had emigrated to England, wrote asking her to join him there. William had been an oboist in the Prussian army, but he had deserted and gone to Bath to continue his musical training and to study astronomy. He needed someone to keep house for him, but his mother disliked the idea of losing Caroline's help; it was only after William promised to send regular funds to provide for a maid to take Caroline's place that her mother released her.

In August of 1772, Caroline and William set out for England, and over the next five years, Caroline's horizons expanded considerably. She learned to speak English, and she studied accounting so that she could manage the household more efficiently. She enlisted a helpful neighbor who taught her English cooking and marketing. William encouraged her to enroll in vocal lessons and to practice on the harpsichord so that she could be a part of his musical performances before small groups. In Caroline's spare time, she and William discussed astronomy, and she began to be interested in the constellations.

English society baffled Caroline. Social life at Bath centered around the spring and autumn flower shows and a great fancy ball. This latter was "so pre-eminent. . . . It is a great subject on which the wondrous female mind, in many instances for months before and after, is absorbed in an ecstatic contemplation" (*London Society* 1870, p. 117).

Such activities seemed "giddy-pated and flighty" to Caroline's German nature. William trained her, much against her will, in the artifice of social exchange affected by the women with whom she came in contact. William also engaged a fashionable dancing mistress, who coached Caroline in proper demeanor and behavior. Her wardrobe came in for extensive revision, and when all this was accomplished, she began to make public appearances at the theater, concerts, and the opera.

Caroline's dream of being self-supporting began to be realized when her demand as a vocalist increased. She sang solo parts in oratorios, which the English greatly admired, and although she refused to perform unless William was there to conduct, by the time she was twenty-seven years old, she was established as a popular vocalist.

William's work in astronomy was onerous by this time, and Caroline's career became subordinate to his requirements for help. She and her brother Alexander undertook the job of copying William's catalogs, tables, and papers that he needed in his astronomical projects.

In those days, telescopes were rare and quite expensive, and they were not as efficient as William might have wished. He used a small Gregorian reflector of about a two-inch aperture, and when he found it difficult to get a reflector with larger dimensions, he attempted to construct one of his own design. He built by hand a magnificent Newtonian telescope of six-foot focal length and began the ambitious undertaking of surveying the entire heavens. When he required additional instruments, Caroline and Alexander worked at the grinding and polishing of specula. His discovery of the planet Uranus, in 1781, changed the character of his professional life, and the Herschels no longer found it necessary to augment their finances by musical performances.

A fortunate sequence of events and encounters led to Herschel's recognition by English scientific circles. One night late in December of 1779, he had very obligingly let an acquaintance look at the moon through one of the telescopes the Herschels had built. The man was entranced and awed by the might and beauty of the view, and he became a strong exponent of Herschel's work. He introduced William to the Royal Society, to court circles, and to King George III, who became William's royal patron.

When Uranus was discovered in 1781, William named the planet "Georgium Sidus" in honor of the king; in appreciation, the regent appointed Herschel to the position of court astronomer the following year. This appointment brought with it an annual salary of 200 pounds a year. Five years later, Caroline was appointed as his assistant with a stipend of 50 pounds annually. Trained as she was in her frugal ways, she survived on this amount and notes in her memoirs: "In October I received twelve pounds ten, being the first quarterly payment of my salary, and the first money I ever in all my lifetime thought myself at liberty to spend to my own liking" (Herschel 1876, p. 76).

Caroline's appointment made her the first woman in England to be honored with such a position in government service. It is likely that never before nor since has any government purchased such a dedicated servant for such a relatively low cost of hire.*

Caroline's efforts were focused on providing a perfect support system for William, and she began rather systematically to collect information and to train herself as competently as possible. She brought a stubborn and powerful determination to studying geometry; collecting formulas, learning logarithmic tables, and learning about the relationship of sidereal time to solar time. She took care of all the laborious

---

* The possible exception was John Flamsteed, another English pioneer in astronomy whom the Crown paid 100 pounds a year from which he was expected to furnish his own instruments. Incidentally, in 1798, Caroline furnished the Royal Society with an Index to Flamsteed's observations, together with a catalogue of 560 stars omitted from his *British Catalogue* and a list of the errata in that publication.

numerical calculations and reductions, all the record keeping, and the other tedious minutiae that required a trained mind but would have consumed too much of Sir William's time. Sister Mary Thomas a Kempis (1955, p. 241) observed of Caroline's labors, "As far as we can learn, not a single computational error has been ascribed to her, and the volume of her work was tremendous."

William's aim was to monitor the skies during every available hour of the night, and when Caroline was not required for other tasks, she swept the skies with a small Newtonian reflector. Her diligence was rewarded by the detection, in early 1783, of such nebulae as Andromeda and Cetus; by the end of that year, she had added fourteen nebulae to the number already cataloged. In an addendum to one of his papers, Sir William gives his sister credit for these discoveries. He also rewarded her with another telescope, one with a small Newtonian sweeper of twenty-seven-inch focal length and a power of 30.

Caroline was the first woman to detect a comet; between the years of 1789 and 1797, she was credited with detecting eight in all.*

One of her greatest services was in arranging catalogs and calculations about 2,500 nebulae based

* Mozans (1913, p. 173) writes of Maria Kirch: "In 1702 . . . she was fortunate enough to discover a comet." He goes on to point out that it was not named for her, however, and most historians give Caroline Herschel the distinction of being the first woman to make such a discovery.

on observations already made. She also reorganized Flamsteed's *British Catalogue* (a listing of nearly 3,000 stars) into zones of one-degree width to enable William to search the skies by a more systematic method.

In 1788 Sir William married the wealthy widow of a London merchant, causing Caroline some apprehension about sharing her brother's affections and household with an English lady. These worries were to prove groundless, however, and Caroline's angst was dispelled by her sister-in-law's gracious and warm disposition. The two became fond and affectionate friends, and Lady Herschel was a source of much support for Caroline in her old age.

Caroline's *Journal* entries recorded visits by English royalty and other celebrities during these years. Her friendship with the Princesses Sophia and Amelia proved to be an absorbing but time-consuming pleasure that she could have easily eschewed in preference for more serious pursuits. Nevertheless, on their visits, the princesses were patiently treated to views of Saturn and its satellites, the mountains of the moon, the variable star Mira Ceti, and Mars. They wanted to know about the rotation of the planets and their satellites, sidereal time; their questioning was vigorous. The Herschels were frequently invited to Windsor, where they spent long evenings discussing their concept of the universe and the solar system's place in space.

Her brother's death on August 25, 1822, caused

Caroline perhaps the deepest hurt of her lifetime. Sir William was a great, good, and kindly man, and he had returned all of Caroline's devotion and affection. His kindness, his charm (he was a superb raconteur), and his almost legendary success as an astronomer brought Caroline delight, and she would listen to no word of praise for her own accomplishments, fearing that his reputation might be diminished if she appeared in a more than secondary role. She was almost pathological in her selflessness where Sir William was concerned; and the teamwork between these two was an extraordinary adjunct that enabled him to achieve a unique position among the world's greatest astronomers.

After her brother's death, Caroline left England and returned to Hanover. She had a small pension, and William had provided a legacy on which she could live. The royal family continued to take an interest in her welfare, even after George IV died and Hanover was separated from the English crown.

But her own work continued. In 1825, she presented the works of Flamsteed to the Royal Academy at Göttingen. To the catalog she had added her own memoranda, which were annotated in the second volume. These were sent to Carl Friedrich Gauss, who thanked her for "these works so precious by themselves but much more so by the numerous enrichments from your own hand, shall always be considered as the greatest ornaments of the library of our Observatory" (Herschel 1876, p. 195).

Caroline spent the last decades of her life in Hanover, organizing and preparing (for her nephew's use) the eight volumes of Sir William's *Book of Sweeps* and the *Catalogue of 2500 Nebulae*. Her nephew, Sir John Herschel, continued the work begun by his father and his indefatigable aunt, and Caroline lived to see Sir John's efforts in mathematics, astronomy, and chemistry added to the already massive labors of the Herschel family.

In 1828 Caroline completed the cataloging of 1,500 nebulae and many star clusters discovered by the Herschels. For this immense and valuable labor, the Royal Astronomical Society voted her a gold medal in recognition of an "extraordinary monument to the unextinguished ardor of a lady of seventy-five in the cause of science." As she intended, this publication was invaluable to her nephew when he began to review his father's work and in the preparation of his own book, *Cape Observations*, a catalog of southern nebulae and star clusters published in 1847.

At the age of eighty-five, Caroline was elected an honorary member of the Royal Astronomical Society. She and Mary Somerville were the first women in England to be honored by this group for their scientific achievements. The Royal Irish Academy also conferred a similar honor on Caroline, and on her ninety-sixth birthday, the King of Prussia awarded her a gold medal of science.

Caroline died at the age of ninety-seven years and ten months. Her epitaph, which she composed herself, reaffirms for all posterity her dedication to science and to her illustrious brother. Even here, she was concerned that Sir William have the greater share of glory.

Here rests the earthly veil of
Caroline Herschel
Born at Hanover, March 16, 1750
Died January 9, 1848

The eyes of her who passed to glory, while below turned to the starry heavens; her own discoveries of the comets and her share in the immortal labours of her Brother, William Herschel, bear witness of this to later ages. The Royal Academy of Dublin and the Royal Astronomical Society of London numbered her among their members. At the age of 97 years and 10 months she fell asleep in happy peace, and in full possession of her faculties; following to a better life her father, Isaac Herschel, who lived to the age of 60 years 7 months and lies buried near this spot since the 25th March, 1767 (*Scripta Mathematica* 21, June 1955, p. 251).

Her patient labors in searching out the reaches of space helped Sir William to become one of the greatest observational astronomers in history. Although she can not be credited with any original work in pure mathematics, there can be no question about the contribution she made in the application of mathematics to advancing the fund of human knowledge.

Along with Emilie de Breteuil, another Frenchwoman won distinction in mathematics in the 1700s. Sophie Germain, who has been called one of the founders of mathematical physics, was born in Paris on April 1, 1776, the daughter of Ambroise François and Marie Germain. She grew up during the impassioned social, economic, and political conflicts of late-eighteenth-century France.

In 1789, Sophie was thirteen years old when the Bastille fell, turning Paris into a bedlam. The streets were filled with discontented Parisians demonstrating

Sophie Germain
1776–1831

their revolutionary sentiments, foraging for food, and reveling in the general anarchy. It was no place for a young girl with Sophie's sensibilities.

Sophie's family was relatively wealthy and could shield her from the revolutionary violence of the streets, but the cost of this protection meant long hours of solitude for the young girl. These hours were spent in her father's ample library, and it was here that she came across the legend of Archimedes' death as recorded by J. E. Montucia's *History of Mathematics.*

It was easy for a lonely young girl to romanticize the fate of Archimedes, killed while absorbed in a geometry problem by a ruthless Carthaginian soldier. She reasoned that if geometry was so very engaging, it must hold wonders worth exploring, and starved as she was for mental stimulation, she was eager to investigate these new wonders.

Her family firmly and stubbornly opposed her decision, but her determination was only strengthened by the vehemence of their opposition. The study of mathematics became a passion for her, one that no amount of familial pressure could smother. Alone and untutored, she went through every book her father's library afforded on the subject.

Her parents, concerned for her health and threatened by the customary wild stories of young girls who were too studious, took desperate measures: They denied her light and heat for her bedroom and confiscated her clothing after she retired at night in order to force her to sleep. Sophie played through this authoritarian charade docilely, but after her parents were in bed, she would wrap herself in quilts, take out a store of hidden candles, and work at her books all night.

After finding her asleep at her desk in the morning, the ink frozen in the ink horn and her slate covered with calculations, her parents finally had the wisdom and grace to relent, freeing Sophie to study and use her genius as she wished. It was a fortunate decision, and Sophie, still without a tutor, spent the years of the Reign of Terror studying differential calculus.

In 1794, the École Polytechnique opened in Paris, and although it did not accept women as students, Sophie very diligently set about collecting the lecture notes of various professors. She found the analysis of J. L. Lagrange particularly interesting to her; following the new revolutionary practice that allowed students to hand in written observations to the professor at the end of the course, she communicated her own composition to Lagrange under the pseudonym of M. le Blanc, the name of a student at the school.

Lagrange was impressed by the paper, and on learning the actual identity of the author, went to her home to praise her as a promising young analyst. Such encouragement from one of the most outstanding mathematicians of the eighteenth century gave Sophie the moral support she lacked at home.

In 1801, Gauss published his *Disquisitiones arith-*

meticae, a masterpiece on the theory of numbers. It was a classic work establishing Gauss's theories of cyclotomy and arithmetical forms, but it was difficult reading, even for experts. The treasures it contained were worthy of study, however, and by 1804, Sophie had become so entranced with Gauss's work that she sent him some of the results of her own mathematical investigations, again using her pseudonym of M. le Blanc.

Gauss was intrigued by Sophie's observations, and the two entered into an extensive correspondence. Sophie did not abandon her disguise, and Gauss did not suspect her identity until 1807 when she became concerned about his safety and interceded on his behalf with the French general, a family friend, who commanded the troops holding Hanover.

General Pernety's troops were besieging Breslau, near Gauss's home. At Sophie's request, the general was kind enough to send an emissary to see that Gauss was safe. It developed that Gauss was fine, but he was confused at the mention of Sophie's name and denied any knowledge of her, because his contacts had been with a M. le Blanc. The misunderstanding was cleared up only after an exchange of letters.

Gauss's correspondence was of interest, for it reflects his liberal views regarding women. These were extraordinary for the times, particularly so for a German man. He wrote (Bell 1937, p. 262):

But how to describe to you my admiration and astonishment at seeing my esteemed correspondent

M. le Blanc metamorphose himself into this illustrious personage [Sophie Germain] who gives such a brilliant example of what I would find it difficult to believe. A taste for the abstract sciences in general and above all the mysteries of numbers is excessively rare; one is astonished at it; the enchanting charms of this sublime science reveal themselves only to those who have the courage to go deeply into it. . . . Indeed nothing could prove to me in so flattering and less equivocal manner that the attractions of this science, which has enriched my life with so many joys, are not chimerical, as the predilection with which you have honored it.

Gauss's letter goes on to discuss points of mathematical interest with Sophie; and he ends his letter with a whimsical touch: "Bronsvic ce 30 Avril 1807 jour de ma naissance" (Brunswick, 30 April 1807, my birthday).

Gauss was deliberate in all he did, and his praise of Sophie was not inspired solely by gratitude for her concern. This was evidenced by another of his letters to a friend, H. W. M. Olbers, on July 21, 1807 (Bell 1937, p. 262):

Lagrange is warmly interested in astronomy and the higher arithmetic; the two test-theorems (for what primes 2 is a cubic or a biquadratic residue), which I also communicated to him some time ago, he considers "among the most beautiful things and the most difficult to prove." But Sophie Germain has sent me the proofs of these; I have not yet been able to go through them, but I believe they are good; at least she had attacked the matter from the right side, only somewhat more diffusely than would be necessary.

Most of Sophie's early research had been in number theory; but around the turn of the century, mathematical interest in Paris began to focus on the work of Ernst Chladni, a German (some writers say Italian; see Coolidge 1951, p. 27) physicist living in Paris. Chladni had studied the vibration of elastic surfaces by sprinkling them with a fine powder, strumming the edges with a bow, then noting the figures formed by the nodal lines.

Chladni's work had sparked considerable interest in the underlying mathematical law concerning vibrations of elastic surfaces. Although a theory for the corresponding problem in one dimension had already been developed, the theory for two dimensions was still too formidable to engage most mathematicians.

Lagrange, who was familiar with the problem, assumed it would yield only to some radically new system of analysis, and when the French Academy of Sciences, by order of Napoleon, offered a prize for the best essay on the mathematical laws of elastic surfaces comparing these laws with experimental data, most mathematicians were reluctant to engage in such a consuming commitment, particularly one that promised so little probability of success.

But Sophie found this project challenging, and whether she was overly confident in her ignorance of the problem's complexity or (what is more likely, considering her later writing—see Todhunter and Pearson 1960, pp. 147–160) because her early training had taught her to cope with discouragement, she undertook the solution of the problem.

By 1811 she was able to submit an anonymous memoir to the Academy, although it was quite evident that her formal training was inadequate for the task she had undertaken. Lagrange, who was a member of the commission evaluating her work, wrote that her method of passing from a line to a surface did not seem to him accurate and complete. Her entry was rejected, but it did not deter her from another try.

In 1813 a second competition was held, and this time Sophie's entry was voted an honorable mention by the commission. The problem continued to interest her, and in 1816 her *Memoir on the Vibrations of Elastic Plates* earned her the prize. (Todhunter [1960, p. 149] writes that the judges were not completely satisfied with her demonstration, and Sophie herself admitted that the agreement between theory and observation was not a rigorous one. Todhunter points out that she "had taken without due examination a formula from a memoir by Euler, *De sono campanarum*, which was incorrect. . . ." She was severely criticized for this and other shortcomings of her memoir by such men as Kirchhoff, Poisson, and even her good friend, Fourier.)

Winning the *grand prix* of the Academy elevated Sophie to the ranks of the most noted mathematicians of the world, and she was welcomed into mathematical circles frequented by many eminent men including A. L. Cauchy, A. M. Ampere, M. H. Navier, A. Legendre, S. D. Poisson, and Fourier. She was feted in a public meeting of the *Institut de France*, and she was invited to attend the sessions of the *Institut*, the highest honor that this famous body had ever conferred on a woman.

Navier was so impressed by Sophie's powers of analysis displayed in her memoirs on vibrating surfaces that he wrote, "It is a work which few men are able to read and which only one woman was able to write" (Mozans 1913, p. 156). Baron de Prony called her the Hypatia of the nineteenth century; and J. Biot wrote, in the *Journal de Savants*, March 1817, that she had probably penetrated the science of mathematics more deeply than any other of her sex, not excepting Mme du Châtelet.

In addition to the memoirs submitted to the Academy, Sophie also published several others dealing with the theory of elasticity. The most important of these included one on the nature, bounds, and extent of elastic surfaces; another on the principles of analysis used in the solution of the problem of elastic surfaces; and one on the curvature of elastic surfaces. This last paper was not published until 1831, after the author's death, and it dealt with mean curvature (defined here as the sum of the reciprocals of the radii of principal curvature; this is often taken as twice the mean curvature), and although Sophie referenced Gauss's work in this area, some writers believe that she failed to grasp how much more important the Gaussian curvature was than that which she treated in her article. (For those who are interested in a critical analysis of this part of Sophie's work, see Todhunter and Pearson 1960, pp. 147–160.)

Despite the importance of the work she did on the theory of elasticity, Sophie is best known for her work in the theory of numbers. Here she demonstrated the impossibility of solving Fermat's last theorem if x, y, and z are not divisible by an odd prime n. (If n is an odd prime < 100, the equation $x^n = y^n = z^n$ is not soluble in integers not divisible by n).

Like her Italian predecessor, Maria Gaetana Agnesi, Sophie Germain was also interested in philosophy. Her paper "Considérations sur l'Etat des Sciences et des Lettres aux Différentes Epoques de Leur Culture" set forth her philosophical views in beautiful literary form.

Sophie also studied chemistry, physics, geography, and history. To each of these fields she brought her own unusual talent and analytical genius, but she continues to be best known for her work in mathematics.

Despite their extensive correspondence, Sophie and Gauss never met, but his respect for her capabilities moved him to recommend to the faculty of the University of Göttingen that Sophie be given an honorary doctor's degree. She died in Paris on June 26, 1831, before the degree could be awarded.

Work was Sophie Germain's whole life; she knew no other, and her last months were spent in the study of mathematics despite the savage pain of a breast cancer that eventually ended her life. Her temperament and disposition were much closer to that of Agnesi than to that of her compatriot, Mme du Châtelet. She chose as her milieu the laboratory and the library, not the salon and the bedroom.

Her portraits show a mature face. There is no flash of coquetry about her, no hint of frivolity or gaiety. One sees, instead, an unmistakable and unalterable honesty and dignity overlaying a quite formidable intelligence.

One of her most sympathetic biographers (Mozans 1913, p. 156) wrote of her:

All things considered, she was probably the most profoundly intellectual woman that France has ever produced. And yet, strange as it may seem, when the state official came to make out the death certificate of this eminent associate and co-worker of the most illustrious members of the French Academy of Sciences, he designated her as a *rentière—annuitant—* not as a *mathématicienne*. Nor is this all. When the Eiffel Tower was erected, in which the engineers were obliged to give special attention to the elasticity of the materials used, there were inscribed on this lofty structure the names of seventy-two savants. But one will not find in this list the name of that daughter of genius, whose researches contributed so much toward establishing the theory of the elasticity of metals— Sophie Germain. Was she excluded from this list for the same reason that Agnesi was ineligible to membership in the French Academy—because she was a woman? It would seem so. If such, indeed, was the case, more is the shame for those who were responsible for such ingratitude toward one who had deserved so well of science, and who by her achievements had won an enviable place in the hall of fame.

The mathematical talents of Mary Fairfax Somerville, called "one of the greatest women scientists England ever produced," might never have been discovered except for a fortunate accident. As was the case in the lives of Newton and Lagrange, her introduction to mathematics was accidental; indeed, chance played a decisive role not only in the discovery of her genius but also in its development.

Until she was around fifteen years old, Mary led the customary, bucolic life of a carefree young girl in eighteenth-century Scotland. She rambled, aimlessly

Mary Fairfax Somerville
1780–1872

and lonely, over the sands and braes of Burntisland, almost totally ignorant that a system of mathematics existed at all. Certainly she never guessed at the impact it was to have on her own life.

Mary Fairfax was born in Jedburgh, Scotland on December 26, 1780. Her father, Sir William Fairfax, was a vice-admiral in the British Navy, and he was frequently away from home for long periods of time. During these absences, the family was forced to live on the strictest economy possible; but although they were poor, both parents boasted that they were of "good birth."

Mary's father was a relative of Lord Fairfax and a descendant of the same ancient and respected Yorkshire family from which George Washington descended. Her mother was Margaret Charters, the daughter of Samuel Charters, the Solicitor of Customs in Scotland, a man of some intellectual pretensions and learning.

According to Mary's own account (Tabor 1933) her mother taught her to read the Bible and say her prayers, but otherwise she was allowed to grow up a "wild creature." Aside from her domestic duties, such as caring for the poultry and the dairy, there was nothing to occupy Mary's time, but she had no playmates and expressed an abhorrence for such childhood pastimes as playing with dolls.

Burntisland was a quiet, quaint little seaport village where women still wore the picturesque costumes described in "Christie Johnstone," and where licensed beggars, or gaberlunzie men, in their blue coats and tin badges, still shuffled through the lanes, repaying the housewife for a handout by whispering the latest bit of gossip in her ear. It was a provincial village where one lived a circumscribed life.

The Fairfax house stood close by the shore, and its large, shaded garden ran down to the sea, hedged by low, dark rocks that were washed by the wild tides and were almost constantly alive with interesting marine life. Many of Mary's lonely young hours were spent exploring along this Scottish seacoast and through the dark moors that also bordered her home; these explorations were to have a lasting effect on her life and interests in nature.

She did not seem to be bookish by inclination, and in fact, by the age of ten, she could scarcely read. Her education had been a rather desultory one, mostly self-directed, quite haphazard and scant. Her father, on his return after a long absence, was shocked to find Mary a "savage" as a result of her carefree life. He promptly sent her away to a fashionable girls' school at Musselburgh.

The school was kept by a Miss Primrose, who had quite uncompromising notions concerning what represented a fitting education for young girls. The rigid discipline there made Mary utterly wretched, and even in her old age, her writings continued to reflect the horror of this experience.

She wrote of the stiff stays with steel busks that were placed in the front of her dress; heavy bands pressed her shoulders back so that her shoulder blades met. A steel rod with a semicircle supporting her chin was clasped to the steel busk in her stays, and in this constrained state, she was expected to prepare her lessons.

One of her first assignments was to learn by rote a page of Johnson's *Dictionary*—to spell the words, give their parts of speech and their meaning, and, as a further exercise in memory, to recall their order of succession. There was little in the banalities taught at the school to stimulate her mind or to make her stay worthwhile, and after a year of study at Musselburgh, Mary returned home only to be reproached for having learned so little. But she was set free once more to enjoy the country life, to study the flowers, birds, and animals, and to absorb the small collection of books that she read in between her domestic duties. Her mother did not particularly mind Mary's reading, but her garrulous Aunt Janet, who had come to live with the family and whose trenchant tongue Mary feared, disapproved vigorously. She is quoted (Tabor 1933, p. 96): "I wonder you let Mary waste her time in reading; she never sews more than if she were a man." Mary's father agreed with Aunt Janet, and as a result Mary was enrolled in a sewing school to learn the tiresome art of stitchery.

Later in life, Mary left a rich account of these early days in letters to her children. She wrote of this period that she was not a favorite of her family nor was she particularly happy herself. She was bored, diffident, somewhat insular, and her life chances did not look at all promising.

An attic window that faced the north afforded her a hidden, cozy place to muse about her problems, and it furnished a vantage point from which to study the stars by night, a pastime that also had a strong influence on her development.

In her boredom, Mary began to teach herself Latin so that she could read Caesar's *Commentaries.* In the summer of her thirteenth year, she met one of her uncles, a Dr. Somerville, who agreed to help her read Virgil. Although he was a good tutor, their association did not prove to be very agreeable, for Mary found that her political views differed from her uncle's more conservative ones; his abuses of the Liberal party were so severe and uncompromising that Mary became a liberal in defiance. She was to credit her uncle's reactionary Tory politics with pushing her toward a more enlightened attitude about women and their education; in later life she often spoke of her sentiments about these injustices and worked actively for women's rights.

Mary's family took an apartment in Edinburgh for a brief period; this move afforded her an opportunity to study arithmetic and writing, the piano, and to continue her study of Latin. But these studies were discontinued when the family returned to Burntisland, and as Mary neared her mid-teens, she began to feel a more acute need for an education. Her family, tradition bound and fearful of the idea, was firmly opposed, and she was forced to spend her time in social activities or in learning domestic arts.

As unlikely as it may seem, it was at a tea party in Burntisland where Mary and a friend were idly leafing through a fashion magazine that she chanced upon some algebraic symbols, which piqued her interest. The story is best told in Mary's own words (Tabor 1933, p. 98):

At the end of the magazine, I read what appeared to me to be simply an arithmetical question, but on turning the page I was surprised to see strange-looking lines mixed with letters, chiefly Xs and Ys, and asked, "What is that?"
"Oh," said the friend, "it's a kind of arithmetic; they call it Algebra; but I can tell you nothing about it."
And we talked about other things; but on going home I thought I would look if any of our books could tell me what was meant by Algebra.

Unfortunately, her home library contained no books on this fascinating new subject; but she did find one on Robertson's *Navigation* that gave her some insight into new and challenging ideas. Parts of the book were beyond her comprehension, but even these passages brought glimmers of promise and reinforced her determination to know more about the strange algebraic symbols and unusual terms she had encountered.

There was no one to whom she could turn for help, however; none of her acquaintances or relatives had any knowledge of science or natural history, and she later observed that even should they have had such knowledge, she would have lacked the courage to seek their help because her efforts would have been ridiculed, she would have been dismissed as a silly and useless dreamer, or worse, there would have been stratagems conceived for confounding her purpose. Chance favored her once again, however, and again it came at an unlikely time and place. She had been sent to learn painting and dancing at Naysmith's Academy, and it was here in a discussion of perspective that Mary overheard the master of the school advise a male student to study Euclid's *Elements of Geometry*, a book the master considered the foundation of perspective and mechanical science. This chance remark gave Mary a clue to the importance of Euclid, but she still faced the unsettling problem of obtaining a copy for her study. It was considered unacceptable for a young girl to walk into a bookseller and ask for a copy of Euclid, and she was too artless or too sensitive to her family's searing and relentless opposition to flaunt tradition further. But despite the crushing restraint she faced from all sides, these two brief chance encounters were to mark the inauspicious beginnings of a self-motivated search for knowledge that would make Mary Fairfax Somerville a leading polymath of her time.

She eventually got her copy of Euclid through her youngest brother's tutor, a Mr. Gaw: Mary happened to be sewing one day in the room where her brother was doing his lessons, and she involuntarily prompted him when he stumbled over the answer to a problem, much to the astonishment of his tutor. This gentleman had assumed that Mary's attentions were completely absorbed by the stitchery in her lap; but upon finding that she had indeed grasped some of the principles of mathematics, he was kind enough to cooperate with her by demonstrating problems in the first book of Euclid. Although the tutor himself was quite limited in mathematical training, Mary did gain enough from his tutelage to continue on her own, memorizing the theorems and repeating them to herself in bed every night.

Her mother was appalled and shamed by such aberrant behavior, and the servants were instructed to confiscate Mary's supply of candles so that she could not study at night. However, by this time, Mary had gone through the first six books of Euclid and, according to Mozans (1913, p. 158), said

I was now thrown on my memory, which I exercised by beginning at the first book and demonstrating in my mind till I could go through the whole. My father came home for a short time, and somehow, or other, finding out what I was about, said to my mother, "Peg, we must put a stop to this, or we shall have Mary in a straight-jacket one of these days." There was X, who went raving mad about the longitude!

Mary's marriage in 1804 to her cousin Samuel Greig gave her a little more freedom to work with mathematics, although this interest did not meet with sympathy from her husband, who had no knowledge of science of any kind and held intellectual women in quite low esteem.

Two sons were born to Mary in this marriage: one died in infancy and the other, Woronzow Greig, later became a barrister-at-law and lived until middle age. Mary lost her husband in 1807, after only three years of married life, and these two deaths coming so close together left her despondent and in rather poor health for several years.

She returned to Burntisland and being financially independent for the first time in her life felt free to begin studying mathematics and astronomy in earnest.

By this time, she had mastered plane and spherical trigonometry, conic sections, and J. Ferguson's *Astronomy* on her own. She had also attempted to study Newton's *Principia*, although she found this latter extremely difficult at the first reading.

The editor of a popular mathematical journal came to her assistance after she had submitted a winning solution to a prize problem published in his journal. (Her first public success appeared in this periodical when she solved a prize problem on Diaphantine equations and was awarded a silver medal cast in her name. There is some question as to whether her entire paper was published at this time, however.)

The editor, to whom she confided her determination to educate herself in mathematics, was kind enough to advise her about a basic course of study, naming for her the classics that she would need to give her a sound background in mathematics. Her delight over these acquisitions was almost pathetic (Tabor 1933, p. 107):

I was thirty-three years of age when I bought this excellent little library. I could hardly believe that I possessed such a treasure when I looked back on the day that I first saw the mysterious word "Algebra," and the long years in which I persevered almost without hope. It taught me never to despair. I had now the means, and pursued my studies with increased assiduity; concealment was no longer possible, nor was it attempted. I was considered eccentric and foolish, and my conduct was highly disapproved by many, especially by some members of my own family. They expected me to entertain and keep a gay house for them, and in that they were disappointed. As I was quite independent, I did not care for their criticism. A great part of the day I was occupied with my children; in the evening I worked.

It was not until her marriage in 1812 to another cousin, William Somerville, that Mary began to realize the extent and the extreme severity of the criticism her studies had inspired. One of her husband's sisters wrote brazenly that she hoped Mary "would give up her foolish manner of life and make a respectable and useful wife" (Tabor 1933, p. 110).

Fortunately, this sentiment was not shared by all members of the family, for William Somerville was an eminently civilized man, a classical scholar, handsome, urbane, and emancipated. He could not have been more supportive of his wife's endeavors, and when she began writing, he helped her to search libraries, read proofs, and check manuscripts.

Dr. Somerville was a surgeon, and, for a while, he was head of the Army medical department. During the early years of their marriage, the couple lived in London and Scotland, and their London home was near the Royal Institution of Great Britain, where Mary could continue her studies.

Dr. Somerville's confidential missions for the British government exposed the couple to a lively intellectual circle. Mary came to know Pierre Laplace, with whom she discussed astronomy and the calculus; she knew Georges Cuvier and Pentland, whose explorations and information helped her later in the preparation of her book on physical geography; and she knew the leading astronomers of England and the continent, one of whom, Sir Edward Parry, named a small island in the Arctic after her. The Somervilles also counted such people as the Napiers (Sir Charles), Caroline and Sir William Herschel, Dr. Whewell, Lord Brougham, and Gay-Lussac among their friends.

In 1826 Mary presented a paper to the Royal Society on "The Magnetic Properties of the Violet Rays of the Solar Spectrum." Although she had begun work on the paper during her first marriage, she had been compelled to work under very limiting conditions considering the complexity of the subject, and her lack of sufficient formal training and an adequately equipped laboratory was reflected in her paper. Although it attracted much interest and attention and was lauded for its ingenuity, the theory it propounded was later very effectively challenged by the research of Moser and Ries.

Dr. Somerville's illness in 1844 sent the couple to Paris, and from then until his death in 1860 they spent most of their time on the continent, returning to England for occasional visits.

Mary's reputation had become established among her friends by this time. In 1827 Lord Brougham, on behalf of the Society for the Diffusion of Useful Knowledge, wrote to Mary's husband inquiring whether Mary could be persuaded to write two volumes, one on Laplace's *Mécanique celeste* and another on Newton's *Principia.*

The last volume of Laplace's *Mécanique celeste* had appeared in 1825, and in it he summarized the work on gravitation of several generations of brilliant mathematicians. Laplace attempted an extensive explanation of the motions of the bodies of the solar system, or, at least, he began to formulate methods for this purpose.

At the time, English scholars had become insular, inspired by a national pride that was a natural outgrowth of their triumph in Newton's advances. Lord Brougham was attempting to break through this insularity and complacency in much the same way Maupertuis and Voltaire had done in France during the previous century. Science (and mathematics, in particular) was at a low ebb in England after the passing of Sir Isaac Newton and the group of scholars associated with him, and some intellectuals saw the need for closer contact with the work of continental scientists and their methods.

Lord Brougham wrote to Dr. Somerville, urging him to exert his influence to persuade his wife to help (Parton 1883, p. 372):

The kind of thing wanted is such a description of that divine work as will both explain to the unlearned the sort of thing it is—the plan, the vast merit, the wonderful truths unfolded or methodized—and the calculus by which all this is accomplished, and will also give a deeper insight to the uninitiated. . . . In England there are now not twenty people who know it even by name. My firm belief is that Mrs. Somerville could add two ciphers to each of those figures. Will you be my counsel in this suit?

Lord Brougham also came in person to press his request that Mary prepare for English readers a popular exposition of these great works, but she was unsure of her qualifications for such a project. She accepted the assignment only on condition that, if she should fail, the manuscript would be destroyed and in the meantime her work would be kept secret. Her apprehensions can be understood when one considers that she was nearing fifty years of age, had very little formal training, had never written for publication, and that the proposed project was a very difficult one indeed.

In addition to all these limitations, her family had

grown (she was now the mother of three daughters), and there were other constraints and demands on her time, which she described in her writings (Tabor 1933, p. 106):

I rose early and made such arrangements with regard to my children and family affairs that I had time to write afterwards; not, however, without many interruptions. A man can always command his time under the plea of business; a woman is not allowed any such excuse. At Chelsea I was always supposed to be at home when friends and acquaintances came out to see me; it would have been unkind not to receive them. Nevertheless, I was sometimes annoyed when in the midst of a difficult problem someone would enter and say, "I have come to spend a few hours with you."

It took all the patience, competence, determination, and organization she could muster to complete the project, but when her work was published, it was far more than the translation Lord Brougham had envisioned in his request. In addition to the views of Laplace, the book also presented Mary's own very valuable independent opinions, and these were accepted by Lord Brougham (and Sir William Herschel) just as they were written.

Mary called her work *The Mechanisms of the Heavens*; it gave a general exposition of the mechanical principles of the universe, the planetary and lunar theories, as well as those of Jupiter's satellites, and other pertinent points. Her translation and commentaries were given in lucid terms so that they could be understood by the average person who had little or no knowledge of mathematics or calculus. This explanatory function was important to Mary, and she said of the scope and purpose of her work (Proctor 1886, p. 5):

A complete acquaintance with physical astronomy can only be attained by those who are well-versed in the highest branches of mathematical and mechanical science; such alone can appreciate the extreme beauty of the results, and the means by which these results are obtained. Nevertheless, a sufficient skill in analysis to follow the general outline, to see the mutual dependence of the several parts of the system, and to comprehend by what means some of the most extraordinary conclusions have been arrived at, is within the reach of many who shrink from the task, appalled by difficulties which perhaps are not more formidable than those incident to the study of the elements of every branch of knowledge, and possibly overrating them by not making a sufficient distinction between the degree of mathematical acquirement necessary for making discoveries and that which is requisite for understanding what others have done. That the study of mathematics, and their application to astronomy, are full of interest, will be allowed by all who have devoted their time and attention to these pursuits; and they only can estimate the delight of arriving at truth, whether it be the discovery of a world or of a new property of numbers."

Mary's translation of Laplace's *Mécanique celeste* was the most famous of her mathematical writings.

After the publication of *The Mechanisms of the Heavens* in 1831, she was raised to the first rank among scientific writers of the time. Laplace made the observation that she was the single woman who understood his work; Poisson expressed doubt that there were twenty people in all of France who could comprehend her writings. Other distinctions were showered on her. The British Royal Society ordered her bust by Chantry to be placed in their great hall; a civil list pension was awarded to her; and her work became a required textbook for the honor students at Cambridge.

Her good friend, Dr. William Whewell, the famous Master of Trinity, wrote from Cambridge, "Mrs. Somerville shows herself in the field in which we mathematicians have been labouring all our lives, and puts us to shame" (Tabor 1933, p. 112).

Mary next turned her mathematical talents to writing *The Connection of the Physical Sciences*, a summary of research into physical phenomena. This book was published in 1834, and several editions were printed. John Couch Adams, the discoverer of Neptune, told her later that a sentence in the book had given him the idea to look for Neptune, whose orbit, mass, and size were all calculated from the unexplained movements of Uranus before the planet itself was found.

Mary also wrote *Physical Geography* (for which she was preached against in York Cathedral) and a number of abstruse monographs on mathematical subjects, one of which was a treatise "On Curves and Surfaces of Higher Orders" that she said later was written "con amore" to fill up her morning hours while spending the winter in southern Italy.

Mary Somerville was primarily a mathematician, and although much of her work was on closely related scientific subjects, she often spoke of the beauty and logic of mathematics with particular feeling. She was bound by her hopes and ambition to use mathematics in a practical sense, for as the American geometer, Julian L. Coolidge, pointed out (1951, p. 25), mathematics was the natural bent of her mind.

The death of her only remaining son in 1860 followed by the death of her husband in 1865 left Mary distraught. She was eighty-one years of age, most of her family was gone, and although her mind was still vigorous and active, her days were lonely and empty. At the suggestion of her daughter, she set to work on a new project.

Molecular and Microscopic Science was published in 1869 when Mary was eighty-nine years old. It was a summary of the most recent discoveries in chemistry and physics. Among her other works were several treatises on subjects related to physics, such as the unpublished *The Form and Rotation of the Earth*. Poisson had encouraged her in this work after publication of *Mechanisms of the Heavens*. She also wrote *The Tides of the Ocean and Atmosphere*.

One might expect that after all her accomplishments

and honors, the discouragement and prejudice that had impeded her in the earlier part of her life would have ended. This was not the case, however. The last twenty-five years of her life were spent in Italy, where she had moved because of her husband's health. This "exile" exposed her to yet another disappointment, which Richard Proctor (1886, p. 13) describes:

It was felt by her friends to be a truly pathetic incident that, of all people in the world, Mrs. Somerville should be debarred the sight of the singular comet of 1834; and the circumstance was symbolical of the whole case of her exile. The only Italian observatory which afforded the necessary implements was in a Jesuit establishment, where no woman was allowed to pass the threshold. At the same hour her heart yearned towards her native Scotland, and her intellect hungered for the congenial intercourse of London; and she looked up at the sky with the mortifying knowledge of what was to be seen there but for the impediment which barred her access to the great telescope at hand. With all her gentleness of temper and her lifelong habit of acquiescence, she suffered deeply, while many of her friends were indignant at the sacrifice.

To summarize Mary Somerville's legacy, one judges that it rests most heavily on her four treatises and her various papers. The first of these papers set forth the solution of a prize problem on Diophantine algebra in 1812. Although she was awarded a silver medal for this solution, it is not known for certain whether this paper was ever published.

Two early papers on her experiments were published and are of some importance. The first was on the magnetizing power of refrangible solar rays and appeared in the *Philosophical Transactions of the Royal Society of London* (vol. 126) in 1836. The second paper was on the transmission of chemical rays of the solar spectrum through different media and appeared in the *Edinburgh Philosophical Journal* (vol. 22) in 1837. Both of these papers were important for the elegance and the methodology of the experimental work, as well as for the lucidity with which it was recorded.

Her *Mechanisms of the Heavens* brought her the widest fame of all her works, but *The Connections of the Physical Sciences*, published in 1834, was also very popular and went through several English and American editions. She stated in the preface of this work that she had as a goal, "to make the laws by which the material world is governed, more familiar to my countrywomen." Whether or not she achieved this aim is problematical, but certainly her work afforded the beginner a much-needed introduction to the physical sciences, for it was written with her customary comprehensible style. At the same time, it was sufficiently authoritative to be relied upon by the most advanced scientists. Some critics called it the best general survey of the physical sciences published in England up to that time.

Physical Geography also had the distinction of being the first important book on this subject in the English language. Mary had followed the advice of Sir John Herschel in the publication of this treatise. Her writing was interrupted by family travels, and Karl Wilhelm von Humboldt's *Kosmos* appeared before her own work was ready. She considered destroying her manuscript, but her husband and Sir John interceded, and it would seem that they were justified in this action for the book went into six editions. It reflects Mary's unusual talent for organization and concentration, as well as her extensive knowledge.

On Molecular and Microscopic Science is the least-known of Mary's works. Its major importance lies in the attempt she made to present the life history of flora and fauna as a triumph of microscopic science.

Mary belonged to a group of scientists who pioneered the effort to arouse English interest in mathematical and scientific progress, and her writings were elegantly free of mathematical intricacies and jargon. Her own early learning difficulties no doubt gave her an insight into the dynamics at work in others that inhibited their understanding of scientific writings.

Mary Somerville's long life was blessed with an extraordinary physical vigor, and her last years were devoted to reading, studying, and writing. At the time of her death she was engaged in several projects, among them a book on quaternions. She was also reviewing a volume, *On the Theory of Differences*. She continued the study of mathematics until the very day of her death at the age of ninety-two.

She was well honored during her lifetime and

afterward. A Victoria gold medal of the Royal Geographical Society was conferred on her in 1869; a similar distinction was awarded her by the Italian Geographical Society. After her death, her name was given to the foundation of Somerville College, one of the five women's colleges now at Oxford. The Mary Somerville scholarship for women in mathematics at Oxford was also established in her honor.

In judging Mary Somerville's work, one is tempted to give quarter because of the odds against her. In a less sentimental attitude, however, it should be noted that there is ample evidence of the disadvantages she encountered because of a lack of rigorous and thorough formal training in mathematics, most specifically in her first work. Although her later writings show her power and thorough mastery of the instruments of mathematical research, they are remarkable less for their scholarship (though this is great) than as an index of what, under more fortunate auspices, she might have produced. Her young, vital, and productive years were wasted.

The stifling prejudice she encountered took a heavy toll, and she was prevented from applying the full powers of her mind, as she herself recognized. Her contemporaries have claimed that almost no department of mathematical research was beyond her powers, and her mental grasp was surpassed by few.

But the rigorous training and scholarly discipline so necessary to original work had been denied her, unfortunately. This is not intended as an apologia for her contributions, however, for Mary Somerville's work needs none. It is mentioned, rather, to give the reader a better insight into the direction her work took and a better understanding of the strength and resilience of this remarkable woman.

Perhaps the most dazzling mathematical genius to surface among women during the past two centuries was the highly gifted Russian, Sonya Corvin-Krukovsky Kovalevsky, born in Moscow on January 15, 1850.* She was destined to become a woman of great

* There is some doubt about her year of birth. Some writers say she was born in 1853. The date above was taken from her biography by her closest friend, Anna Carlotta, Duchess of Cajanello (Leffler 1895). Her biographers have also used a variety of ways to spell her name. Again, the form used here is that employed by Anna Carlotta.

Sonya Corvin-Krukovsky Kovalevsky
1850–1891

strengths shadowed by great vulnerabilities and the imprint she left on mathematics promises to be an enduring one.

Sonya was born into an authoritarian, patriarchal family. Her father was a proud, disciplined man whose displeasure could throw the entire Krukovsky household into a paralytic terror. A general in the Russian army, he was required to move his family frequently in accordance with the requirements of his military duties. When Sonya was around six years of age, General Krukovsky retired, and the family went to settle on their prosperous estate at Palibino. The Krukovskys lived in a remote part of Russia, near the Lithuanian border, where wolves howled about the house on cold winter nights and where there were endless, placid hours for a curious child to fill.

Sonya (or Sophia as she was sometimes called) has left an enchanting written account (Leffler 1895) based on her childhood recollections of life at Palibino. These childhood memories, anchored in wistfulness and nostalgia, furnish the reader with a significant insight into the complex nature of her personality and the psychological dynamics that shaped her.

She was the middle child of three, and when she was still quite young, she reached the conviction (true or not) that her adored older sister Anuita and her brother Feyda eclipsed her in their parents' affections. Impressions of this deep hurt were fixed and exact in Sonya's memory, and they were to fill her with a numbing self-doubt that crippled and embittered her for the balance of her life.

Despite Sonya's authoritarian upbringing (or perhaps because of it), she was surprisingly undisciplined at times. She had a strong-willed nature and on occasion was given to extravagant affection and astonishing jealousy. Her extraordinary individuality, her tensions, and her whims made it difficult for her to live in harmony with others. She very often required a devotion from friends that was beyond human capacity, and her caprice was often burdensome even for those who loved her.

There was a family tradition of mathematical talent in Sonya's background. Those who believe in the theory of "mathematical genes" might doubt that she began *ab initio* in building her own mathematical repertoire. She may have inherited a legacy of talent from her grandfather Feodor Feodorovitch Schubert, who was a fine mathematician and was at one time head of the Infantry Topographical Corps in the Russian army. His father, Sonya's great-grandfather, was even more noted as a mathematician and as an astronomer.

In her book, *Recollections of Childhood*, Sonya tells also of long philosophical discussions she had with her uncle Piotr, who cherished a profound respect for mathematics and was able to transmit this reverence to Sonya, although he was not formally trained in the subject himself. She wrote of hearing from this uncle about such ideas as "the quadrature of the circle, about the asymptotes which the curve always approaches without ever attaining them, and about many other things of the same sort—the sense of which I could not of course understand as yet; but which acted on my inspiration" (Leffler 1895, p. 65). Another factor that may have attracted her to the study of mathematics was a rather singular wallpaper that had been used in one of the children's rooms at Palibino. It seems that not enough paper had been sent out from St. Petersburg to repaper all of the rooms at the old estate, and one room was left unfinished until more paper arrived. During this interim, the room was partially covered with an old paper left from an earlier time. Sonya describes the influence of this old paper (Leffler 1895, p. 66):

By a happy accident the paper used for this first covering consisted of sheets of Ostrogradsky's lithographed lectures on the differential and the integral calculus, bought by my father in his youth.

These sheets, spotted over with strange incomprehensible formulae, soon attracted my attention. I remember how, in my childhood, I passed whole hours before that mysterious wall, trying to decipher even a single phrase, and to discover the order in which the sheets ought to follow each other. By dint of prolonged and daily scrutiny, the external aspect of many among these formulae was fairly engraved on my memory, and even the text left a deep trace on my brain, although at the moment of reading it was incomprehensible to me.

When, many years later, as a girl of fifteen, I took my first lesson in differential calculus from the famous teacher in mathematics in Petersburg, Alexander Nikolaevitch Strannoliubsky, he was astonished at the quickness with which I grasped and assimilated the conceptions of the terms and derivatives, "just as if I had known them before." I remember that this was precisely the way in which he expressed himself, and in truth the fact was that at the moment when he began to explain to me these conceptions, I immediately and vividly remembered that all this had stood on the pages of Ostrogradsky, so memorable to me, and the conception of space seemed to have been familiar to me for a long time."

Sonya was fully as gifted in her literary talents as in mathematics, and she wavered between these fields. She eventually tried her hand at both. Her sister Anuita, while still in her teens, published a short story in a popular magazine edited by Fyodor Dostoevsky, thus beginning a close friendship with the author, who introduced Sonya and Anuita into an elite circle of European intellectuals living in Moscow.

Although her father had reluctantly allowed Sonya to study mathematics with Strannoliubsky at the naval school in St. Petersburg, he was unsettled at the prospect of her following this uncommon pursuit as a serious career. In addition to his objections, there was also another serious impediment to her study: Russian universities were closed to women students, and Sonya's first shy hints about studying in a foreign university were met with stern lectures from her father about such improprieties in young girls.

At the time, Russia was experiencing an increasingly massive generation gap, particularly in the homes of the aristocracy, where the literate young people were becoming rebellious, filled with absolutes, sagacity, and daring. Bewildered parents, confronted with the heterodoxies of their young, became hostile and even more authoritarian than before, and in a vicious circle, the young devised ever more devious ways for escaping parental pressures.

Both Sonya and Anuita were caught up in this conflict between the generations, and the device they settled on was a popular one, for the same situation plagued hundreds of young girls all across Russia. A convenient means of escape for those who could manage it was for one of the girls to contract a nominal, platonic marriage with an acceptable young man for the sole purpose of gaining the freedom to travel. The wife then could go (without public censure) to a foreign university and her sister, or girl friends, could accompany her with complete respectability.

With this scheme in mind, Sonya and Anuita reviewed their coterie of friends and acquaintances and found a likely prospect in Vladimir Kovalevsky, whom Sonya insisted on marrying over her parents' objections. Kovalevsky was a student of paleontology at Moscow University, and he was an apt choice for he was much impressed with Sonya's mathematical talents, her fluency in languages, her literary efforts, and her remarkable beauty. He was quite agreeable to the girls' plans, and in the fall of 1868, he and Sonya were married. The following spring, the couple went to live in Heidelberg, where Sonya was able to attend school and hear the most eminent professors at the famous old university, the oldest and most respected in Germany. Here she heard Leo Königsberger and Emil Du Bois-Reymond lecture on mathematics and Gustav R. Kirchhoff and Hermann L. F. von Helmholtz lecture on physics.

It was quite unusual for one so young and beautiful to be interested in mathematics and the sciences, and almost from the outset, the German professors were impressed with both Sonya's ability and her demeanor. She had a natural shyness that bordered on awkwardness, and the Germans found this quality most appealing. Even the townspeople heard of her and her capabilities, and they came to point her out in the streets. But all of this attention left Sonya essentially unchanged; it was an implacable part of her nature that no matter how celebrated she became she was never to lose the basic insecurities that she had acquired in childhood.

Königsberger, one of Sonya's favorite teachers at Heidelberg, was a former student of Karl T. Weierstrass, a logician whose reputation and influence were enormous among European scholars at the time. After two years of study with Königsberger, Sonya caught his enthusiasm for the master, and she decided to make an effort to study with the famous Weierstrass. When she arrived in Berlin for the fall term in August 1870, however, she found that the university would not accept women and would make no exception in her case despite the favorable recommendations she had brought from her former professors.

With the university doors closed to her, Sonya made a direct appeal to Weierstrass himself for help with private lessons. The celebrated "father of mathematical analysis" was incredulous at her request, but he was a sympathetic and understanding man, open to his students, and never very much bothered with a "great man" complex. Although he was in his fifties by this time, he could recall vividly how in his own ambitious youth, Christoph Gudermann had very kindly taken him on as a student and had given him guidance toward becoming a mathematician. Memory of this old favor disposed him to listen to Sonya's request.

Sonya was awed by the renowned Weierstrass. After all, she was barely out of her teens, had led a relatively sheltered life, and she was frightened of both failure and her own audacity. But she was earnest, eager and determined.

Perhaps more to get rid of her than for any other purpose, Weierstrass gave her a set of problems he had prepared for some of his more advanced students to solve. To his great astonishment, she not only solved them rapidly, but she came up with solutions that were clear and original. Her earnestness and her cleverness impressed the master favorably, and he wrote to Königsberger asking about her and her mathematical aptitudes. He inquired specifically whether "the lady's personality offers the necessary guarantees" (Bell 1965, p. 425).

Of course, Königsberger was able to vouch for both her respectability and her scholarship, and, after receiving these assurances, Weierstrass attempted to get the university senate to admit Sonya to his lectures. He was brusquely refused, but the kindly Weierstrass agreed to make his lectures available to her anyway, and for the next four years, she was his pupil. He shared his lecture notes, his unpublished works, the latest scientific developments, and the new theories of geometry with her. He was by far the most influential teacher in her life.

One of Sonya's gravest indiscretions in her relationship with Weierstrass was her failure to explain to him in the beginning about her marital arrangement. Her nominal marriage was not always sufficient to shield her from criticism, and, indeed, R. W. Bunsen, the famous chemist whom Sonya had met while she was still at Heidelberg, warned Weierstrass once that she was "a dangerous woman." Fortunately, Weierstrass could laugh off Bunsen's intemperate judgment, since at the time, Sonya had already been his student for two years.

The story Bunsen told had an interesting background, however. He was an irascible old bachelor, a woman hater, and Sonya had outwitted him once, an embittering experience for him. He prided himself that his laboratory was a totally masculine scene, untarnished by women, and the crotchety chemist aimed to keep it that way, particularly where Russian women were concerned. One of Sonya's Russian girlfriends had deeply wanted to study chemistry with Bunsen, and after being thrown out of his laboratory, she had appealed to Sonya for help. Bunsen, despite his crusty nature, was no match for Sonya's persuasive powers, and somehow she beguiled him into admitting her friend as a student in his precious laboratory. It was not until after Sonya had left that Bunsen recognized how he had been manipulated, and he sought his revenge by viciously assailing Sonya's good name. As he told Weierstrass later, "And now *that woman* has made me eat my own words" (Bell 1965, p. 425).

During the four years of study with Weierstrass, Sonya completed the university course of mathematics and wrote several important papers. Her doctoral dissertation, "On the Theory of Partial Differential Equations," dealt with a rather general system of differential equations of the first order in any number of variables. Weierstrass had already given an analogous structure for total equations; Sonya's paper extended this to partial differential equations.

She also published "On the Reduction of a Definite Class of Abelian Integrals of the Third Range," again building on Weierstrass' earlier paper on the theory of Abelian integrals.

Her other publications include a paper entitled

"Supplementary Research and Observations on La-place's Research on the Form of the Saturn Ring," and another paper, "On the Property of a System of Equations." In 1874 Sonya was granted her doctorate from the University of Göttingen, and by special dispensation she was exempted from the *viva voce* examination, since, as she put it "I do not know whether I have sufficient aplomb to undergo an *examen rigorosum* . . . having to face men with whom I am altogether unacquainted would confuse me" (Leffler 1895, p. 160). The merit of her written work and testimonials of the scientists with whom she had worked earned her this rare exemption.

After the long years of severe study with Weier-strass, Sonya returned to Russia to relax with friends and relatives at Palibino, St. Petersburg, and Moscow. Weierstrass had attempted to find a position for her that was worthy of her talents, but his efforts had been rebuffed, leaving him disgusted with the bigoted and orthodox mentality of academic cliques.

Finding no market for her mathematical training, Sonya began building another kind of life for herself. Her husband was by now professor of paleontology at Moscow University, where Sonya spent much of her time with an interesting group of friends and relatives. Russian intellectuals welcomed her back into their community, and she busied herself with writing newspaper articles, poetry, theatrical crit-icisms, and a small novel, *The Privat-Docent*, which was considered to show great promise. Her struggle for an education had made her a strong advocate of women's rights, and much of her literary work centered on this theme.

Her only child, Fufa (or Foufie as her mother called her) was born in October 1878, and later that year, Sonya wrote Weierstrass, whom she had neglected for almost three years, that she was anxious to return to her work in mathematics. Her husband, by now famous for his own scientific writings, had become involved with several unfortunate business enterprises, and his losses had put a severe strain on their fragile marriage. Sonya had a small income from her father's estate, but it was not sufficient to sustain her and her daughter, and she was looking for a position that would be adequate for this.

Impatient to be back at work, Sonya left for Berlin on her own. Adversity had forced her to the conclusion that she was a born mathematician, and this was the path she would be forced by her nature to follow. Weierstrass was kind enough to guide her into work on the refraction of light in a crystalline medium, and at the scientific congress in Odessa in September 1883, she gave a paper on the results of her research. Although the paper was well received, it reflected Sonya's long vacation away from scientific work; many years later Vito Volterra, an Italian mathematician, pointed out an error that had escaped the notice of both Weierstrass and Sonya. (Weierstrass was seventy years old at the time and was "brain-weary," as he put it.)

The next few years of Sonya's life were quiet ones spent in Paris, where she became close friends with a handsome young Pole. He was a mathematician, a revolutionary, and a poet; for a time, this idyllic friendship was one of the closest Sonya was ever to have. Their common interests brought a responsive interchange of thought, and Sonya, who was capable of monumental emotions, was intoxicated by this union.

The bond that united her to her husband had always been tenuous, and it had become even more strained because of his economic reverses. It was not completely broken however, as she was to learn. She was distraught at the news of Vladimir's tragic death in the spring of 1883. He had taken his own life, and Sonya reproached herself relentlessly for not remaining in Moscow with him, although by doing so she would surely have been committing herself to an almost unbearable life.

Nevertheless, her grief was massive and bitter. The shock prostrated her, and she shut herself up alone without food for four days. She finally lost consciousness and when she recovered asked for pencil and paper so that she could lose herself in scribbling mathematical formulas. One of her friends wrote (in the flowery prose of the last century) that after the shock of Vladimir's death, Sonya lost "the freshness of youth . . . [and] her complexion, and a deep furrow, nevermore to be effaced, was drawn by care across her brow" (Leffler 1895, p. 202).

During her stay in St. Petersburg in 1876, Sonya had made friends with Gosta Mittag-Leffler, who had also been a pupil of Weierstrass. Leffler had since become professor of mathematics in the new University of Stockholm, and one of his first steps was to induce the authorities there to appoint "Fru" Kovalevsky as privatdocent. He had not only been impressed with Sonya's acuteness and perception but he had also taken a warm interest in the "woman question" and was eager that the new university be the first to attract a great woman mathematician. Earlier in the century, Arthur Cayley had attempted to open up the mathematics department at Cambridge for women, but his efforts had failed for many of the same reasons Weierstrass' had failed in Berlin when he had tried to intercede on Sonya's behalf. The Swedish authorities were more enlightened, however, and in November 1883, Sonya left for Stockholm, where she was to lecture on the theory of partial differential equations.

Sonya was welcomed into the Mittag-Leffler family and became close friends with Gosta's sister, Anna Carlotta Leffler, a writer with whom Sonya was later to make a solemn pact. If either should die, the other would write a biography of the deceased. Anna Carlotta (later the Duchess of Cajanello) kept her word; and after Sonya's death, Anna wrote a most sympathetic biography of her friend's life.

Out of necessity and ambition, Sonya had learned several languages. Although she seemed to have an unusual talent in this direction there were limitations, and she found it difficult to express her thoughts precisely in languages other than Russian. Her course of lectures at Stockholm were given in German, a language she spoke, if imperfectly, still comprehensibly, and she was quite popular with her students. At the end of the term, they gave her an elaborately framed photograph of the class and a very warm and enthusiastic speech lauding her efforts.

Her capacity as a mathematician may be judged by examining the list of her course titles. They included such subjects as: the theory of derived partial equations; the theory of Abelian functions according to Weierstrass; curves defined by differential equations according to Poincaré; the theory of potential functions; applications of the theory of elliptic functions; application of analysis to the theory of whole numbers, and so forth.

Her friend Mittag-Leffler raised private funds to give Sonya an official appointment at the university. Several donors pledged a small sum that the university matched so that Sonya earned a scant living for herself and Foufie. There continued to be conservative opposition to the employment of a woman as a university professor, however, and it was not until five years after she began teaching at the university that Mittag-Leffler was able to obtain for her a tenured appointment, which she was to enjoy only one year.

The high point of Sonya's career came on Christmas Eve of 1888, when she was presented the famous *Prix Bordin* of the French Academy of Sciences in recognition of her winning memoir *On the Problem of the Rotation of a Solid Body about a Fixed Point.* The rules of competition for such prizes dictated that each entry be submitted anonymously. The author's name was sealed into an envelope bearing the same motto as that inscribed on the memoir, and the envelope was not to be opened until after the competing work won the prize. Such a process forestalled any suggestion of favoritism or influence on the part of the judges; when the jury of the Academy chose Sonya's entry, it was in utter ignorance that the winner was a woman.

The excellence of her entry was judged to be so exceptional that the value of the prize was increased from the previously announced 3,000 francs to 5,000 francs "on account of the quite extraordinary service rendered to mathematical physics by this work" (Mozans 1913, p. 164).

Incidentally, the motto on Sonya's prize-winning essay was "Say what you know, do what you must, come what may." By order of the Academy, her work is published in full in the *Mémoires présentés à* l'Académie (*Mémoires des Savants étrangers*), Volume 31, 1884.

Up until this work by Sonya, only two cases had been found that gave the complete solution of the differential equations involved in the rotation of a solid body about a fixed point, but her memoir discussed a totally new case in which the complete solution was exhaustively advanced. The underlying basis of her memoir was an extension of the ideas of Weierstrass and his work on ultraelliptic integrals, but her extension solved a problem that had long baffled mathematicians.

In the following year, the Stockholm Academy also honored Sonya with a prize of 1,500 kroner for two more works built on her original essay. All of this public attention finally moved the University of Stockholm to offer Sonya the coveted professorship.

It was not until December 2, 1889, that Sonya received her first formal recognition from Russian academic circles. She became the first woman Corresponding Member of the Russian Academy of Sciences, an honor she found most encouraging. By this time, Stockholm had begun to depress her, and she longed for an offer to teach in a Russian university, but despite her election to the Russian Academy of Sciences, no teaching position was offered her.

Though Sonya was undeniably a gifted mathematician and had a very practical side to her personality, many of her reactions and judgments were intensely personal. She had a poet's interest in people and their feelings and called herself Bohemian by nature. She attributed this trait to one of her ancestors, a gypsy great-grandmother, whose marriage to Sonya's grandfather had cost him his claim to the title of "prince." She also claimed to have inherited a certain prescience and once wrote to Anna Carlotta (Leffler 1895, p. 270):

It is strange, but the longer I live the more I am governed by the feeling of fatalism, or rather predestination. The feeling of free will, said to be innate in man, fails me more and more. I feel so deeply that however much I may struggle, I cannot change fate one jot. I am now almost resigned. I work because I feel I am at the worst. I can neither wish nor hope for anything. You have no idea how indifferent I am to everything.

This dark side of her personality was strong in Sonya, and she believed earnestly in dreams as portents, in forebodings, and in revelations of other kinds. As she grew older, she predicted that certain years would be "lucky" or "good" years; and, whether by chance or self-fulfilling prophecy, she was often correct in these predictions.

During the months she spent preparing her essay for the *Prix Bordin*, Sonya expended great currents of passion and energy. Of necessity, most of her work on the memoir was done at night, but in addition to the burden of this creative work, her teaching position, her social responsibilities to her sister, and the care of her child, she also had a far more absorbing and electric distraction, for she was once again rapturously in love.

The man who evoked this keen and powerful passion in Sonya is never identified in any of her writings, though he is referred to as Maxim in one source. Whoever he was, it is apparent that he returned her devotion. He was understanding, kind, genial, anxious only about her well being, and willing to make whatever compromises were necessary on her behalf. It was Sonya's own inflexibility that kept the two apart, for she was never able to compromise between her work and her personal life and felt with bitterness and irony that she should sacrifice either her ambition or her love. To withdraw from the competition in Paris would have represented in her mind (and to the world, she thought) a conspicuous proof of woman's incompetence. The strong force of circumstances propelled her into a frenetic effort to win a prize, the enjoyment of which was diminished because of her dual loyalties.

Her childish self-doubts also forced her to make unreasonable and tyrannical demands on her Maxim, and devoted though he was, the impossibility of reconciling her difficult claims finally became too much for both of them, and in the end, the love affair gradually foundered and died.

Sonya attempted to hide from her less intimate friends the subsequent heartbreak she suffered. She preferred instead to express her feelings in literary compositions. In 1889 her novel *The Rayevsky Sisters* was published. The book was an account of her childhood recollections, and it was generously praised by literary critics. Written in Russian, the novel was published in Swedish and then in Danish, and critics judged that it "equalled the best writers of Russian literature in style and content" (Bell 1965, p. 423).

Sonya also published a lesser-known novel, *Vera Vorontzoff*, depicting life in Russia, and to those who expressed surprise at her versatility, she once answered (Leffler 1895, p. 317):

I understand your surprise at my being able to busy myself simultaneously with literature and mathematics. Many who have never had an opportunity of knowing any more about mathematics confound it with arithmetic, and consider it an arid science. In reality, however, it is a science which requires a great amount of imagination, and one of the leading mathematicians of our century states the case quite correctly when he says that it is impossible to be a mathematician without being a poet in soul. Only, of course, in order to comprehend the accuracy of this definition, one must renounce the ancient prejudice that a poet must invent something which does not exist, that imagination and invention are identical. It seems to me that the poet has only to perceive that which others do not perceive, to look deeper than others look. And the mathematician must do the same thing. As for myself, all my life I have been unable to decide for which I had the greater inclination, mathematics or literature. As soon as my brain grows wearied of purely abstract speculations it immediately begins to incline to observations on life, to narrative, and vice versa, everything in life begins to appear insignificant and uninteresting, and only the eternal, immutable laws of science attract me. It is very possible that I should have accomplished more in either of these lines, if I had devoted myself exclusively to it; nevertheless, I cannot give up either of them completely.

After her love affair with Maxim had ended, the remaining months of her life were troubled ones. Her beloved sister Anuita was dying a slow, painful death in Moscow, and Sonya's free time was often spent in making the tiresome, ugly journey back and forth from Stockholm. She was separated from Foufie, who was left behind in Moscow, and although Sonya was only forty-one years old at the time and had talked with friends about new plans for work in mathematics and in literature, there is little doubt that she spent her last months in a severely dispirited, hopeless mood.

Her last journey between the two cities in February 1891 was a particularly trying one. She was preoccupied with worry, anxious and discouraged about her family and her own future, and had little time to think about the exigencies of travel. As a result, she found herself caught in the middle of the night at cold, deserted stations where she was forced to struggle alone with her heavy luggage. Bone tired and frozen, she simply made too many demands on her already diminished energies, and before she reached Stockholm, she became feverish and ill with influenza, which was epidemic at the time. Her death a few days later came as a shock to her close friends and to the mathematical world.

Although she most surely would have preferred being returned to Russia, Sonya was buried in Stockholm. Her devoted Swedish friends thought of her as one of their own, and they were determined to preserve her memory well.

It is interesting to note that this shy daughter of Russia had become a public figure, and not even her tomb could coffer her from the curious. Four years after her death, her brain, which had been preserved in alcohol, was weighed and compared to the weight of Hermann von Helmholtz's brain. Professor Gustaf Retzius wrote an elaborate account of this procedure and estimated that if the body weights of these two eminent people were considered, the relative amount of brain tissue was greater in the woman than in the man. This statement brought to ignominious failure yet another attempt to show that man's brain was superior to that of woman.

Russia has been one of the most generous countries in representing mathematicians on postage stamps, and after her death, Sonya was honored by having a commemorative stamp issued in her name. She is the only woman in this field to have been so honored.

Sonya was perhaps one of the most glamorous

women to win acclaim in the field of mathematics. She was also one of the most impressive, because her talent and work were of a superlative rank. She was not content, as many other prominent women in mathematics have been, to use her talents in translating the work of others. Her discoveries derived from her extraordinary abilities for mathematical research; although she was first and foremost a disciple of Weierstress and was concerned to prove, through her work, the power of his philosophy, her own efforts were sufficiently creative to merit considerable respect.

Her friendship with Weierstrass continued throughout her life, and their voluminous correspondence would have been of considerable scientific interest and importance had it been preserved. Unfortunately, after Sonya's death, Weierstrass burned her letters, and her own papers were left in an untidy disarray, fragmented, and in utter confusion.

As Weierstrass' pupil, Sonya concentrated on the field of analysis and the application of the techniques of analysis to questions in mathematical physics. Infinite series, called the keynote of the nineteenth-century mathematical renaissance, were a central concern of both Weierstrass and Sonya. She is occasionally given specific recognition for "Kovalevsky's theorem" in discussions of Cauchy's problem. This general problem concerns second-order linear partial differential equations in one dependent and *n* independent variables; Sonya's theorem helped to establish rigorously the first existence theorem associated with the question. H. Meschkowski (1964, p. 86) has written of the efforts and effect of the Weierstrass school:

In modern textbooks on function theory there are many theorems and methods of proof which we owe to Weierstrass. Of no less importance for twentieth-century mathematics, however, is the "arithmetization of analysis" he and his school introduced. In the notebooks of students of Karl Weierstrass we no longer find those dubious "infinitely small quantities" which were common in the textbooks of the nineteenth century. Weierstrass and his students "arithmetized" analysis; they reduced statements about limits to equations or inequalities between rational numbers.

Sonya was an effective part of this school, and although her scientific life was brief, it was brilliant, and the mathematical world owes her more than a passing reference.

Emmy Noether's personality and life pattern re-
sembled in some ways those of her French predecessor, Sophie Germain. As Sophie had done in the
previous century, Emmy Noether filled her life with
mathematics, and if politics had not obtruded, she
would have been content to invest all her hours and
energies in this pursuit. She had neither the deep
emotional needs nor Bohemian tendencies of Mme
du Châtelet; nor did she have the self-doubts and
insecurities of Sonya Kovalevsky. Hers was a more
placid, studious nature.

Emmy (Amalie) Noether
1882–1935

Emmy (or Amalie as she was named) was born on March 23, 1882, in Erlangen, a small university town lying in the fertile plains of southern Germany. Her father, Max Noether, a professor at the University of Erlangen, was already distinguished as a great mathematician, having played an important role in the development of the theory of algebraic functions.

Emmy was trained in the household arts and the feminine graces thought so essential for a young girl. She cooked and cleaned, shopped in the little town, and went to dances where she flirted awkwardly with the boys from the university.

Her nature was not a rebellious one. She was described (Weyl 1935, p. 205) as "warm like a loaf of bread. . . . There irradiated from her a broad, comforting, vital warmth." If the ambience of her home had been different, she might never have chosen a career in mathematics, but the provocative discussions that swooped and soared around the young Emmy's head sparked an interest that was overpowering.

Max Noether was a strong influence on the early thinking of his children. Both Emmy and her younger brother Fritz followed in their father's profession, and the Noether family has been cited as yet another example of the phenomenon of the hereditary nature of mathematical talent, the most shining exemplar of which was the Swiss dynasty of the Bernoullis, whose ten members were prominent in mathematics for over three generations.

Emmy was tutored by Paul Gordon, a family friend who taught at the university. Although her later mathematical interests were not wholly consonant with those of Gordon, in 1907 Emmy wrote her doctoral thesis, "On Complete Systems of Invariants for Ternary Biquadratic Forms," under Gordon's tutelage. Her thesis was called an awe-inspiring work, but Emmy dismissed it later as a "jungle of formulas" (Reid 1970, p. 166). She continued to have a deep reverence for Gordon, however, and Hermann Weyl, the renowned mathematician who was a friend and colleague of Emmy's, wrote after her death that Gordon's photograph decorated the wall of her study at Göttingen for many years.

After Gordon's retirement, Emmy's achievements began to reflect her larger talent for conceptual axiomatic thinking, and she moved away from Gordon's formalist approach. During these years she was tutored by two other algebraists, Ernst Fischer and Erhard Schmidt, and her study focused primarily on finite rationals and integral bases. During this time, she was also asked to lecture at the university, occasionally substituting for her father when he was ill.

With a change in her family situation, Emmy was persuaded to move to Göttingen. Her father had retired, her mother had recently died, and her brother Fritz, who had been a student of mathematics at Göttingen, was now in the Army. Her own interests were close to the work being done by David Hilbert at Göttingen, and on one of her visits Hilbert persuaded her to remain. He and Felix Klein were working on the general theory of relativity, and Emmy with her theoretic knowledge of invariants was useful and welcome to work with them. Here Emmy became a part of one of the most creative circles of research in postwar Göttingen, and Weyl (1935, p. 207) wrote that "for two of the most significant sides of the theory of relativity, she gave at that time the genuine and universal mathematical formulation."

It was also here that Emmy became interested in establishing, on an axiomatic basis, a completely general theory of ideals, and her work at Göttingen was to contribute toward making the axiomatic method a powerful instrument of mathematical research.

Although Göttingen was the first German university to grant a doctoral degree to a female, there was still considerable opposition to granting "habilitation" to women. Emmy Noether was no exception, and despite her qualifications, no formal appointment as an academic lecturer was proffered. Hilbert attempted to remedy this injustice by petitioning the Philosophical faculty for Emmy's habilitation, but his efforts met with failure because some members of the faculty were implacably opposed to admitting women as lecturers. The Philosophical faculty included philosophers, philologists, and historians in addition to natural scientists and mathematicians. The nonmathematical members argued (Reid 1970, p. 143): "How can it be allowed that a woman become a Privatdozent? Having become a Privatdozent, she can then become a professor and a member of the University Senate. . . . What will our soldiers think when they return to the University and find that they are expected to learn at the feet of a woman?"

Such atavism annoyed Hilbert, and he further alienated some of the faculty by declaring at one of the meetings (Reid 1970, p. 143): "Meine Herren, I do not see that the sex of the candidate is an argument against her admission as a Privatdozent. After all, the Senate is not a bathhouse."

As a device for frustrating the countervailing members of the faculty, Emmy and Hilbert worked out a system so that Emmy could give the lectures, though these continued to be announced under Hilbert's name. It was not until 1919, after World War I had ended and the proclamation of the German Republic had changed the social climate, that her habilitation finally became possible.

In 1922, Emmy was nominated to the position of a "nichtbeamteter ausserordentlicher Professor," but despite the impressive title, this appointment carried with it little obligation and no salary. She was also honored with a "Lehrauftrag" in algebra; this appointment brought her a very modest remuneration.

It is doubtful that Emmy's true genius was appreciated very widely until 1920 when she coauthored a paper on differential operators, which gave evidence that she had evolved into a truly great mathematician.* This paper marked a decisive turning point for her work, revealing for the first time her strong interest in the conceptual axiomatic approach and establishing Emmy Noether as a force in altering the face of algebra.

This branch of knowledge had changed considerably in the nearly 1,500 years that intervened between Hypatia's *Diophantine Analysis* and Emmy Noether's day. Where Diophantus and Hypatia had been concerned with equation solving and the results of algebraic operations, modern practitioners were more concerned with the formal properties of the algebraic operations. Twentieth-century abstract algebraists were studying commutativity, associativity, distributivity; they were investigating the mathematical systems that result if one of these laws is not assumed; and they were generalizing the number system to other systems, which had been named *fields, rings, groups, near rings,* and so forth.

This axiomatic trend had been accelerated by the work of such people as Leopold Kronecker, G. Peano, K. Weierstrass and S. Kovalevsky, J. W. R. Dedekind, and D. Hilbert, among others. The appearance of Bertrand Russell's *Principia Mathematica* in 1910–1913 introduced new controversies and

* She was thirty-eight years old at the time. Such a late development is a rare phenomenon among mathematicians. In most instances, creative talent is more forceful in early youth.

added totally new logics to the vast domain of mathematics. Just as Newton's *Principia* had established a novel way of thinking for decades after its publication, so did Russell's *Principia* help to establish research into symbolic logic as a major new branch of mathematics. The enormously expanded role of the axiomatic approach became perhaps the most remarkable aspect of twentieth-century mathematics.

Emmy Noether was close to the development of this trend. She had built on her father's work, particularly his residual theorem; during the 1920s, she fitted this theorem into her general theory of ideals in arbitrary rings, helping to further establish the axiomatic and integrative tendencies of abstract algebras.

During the late 1920s, she began investigating the structure of the noncommutative algebras, their representations by linear transformations, and their application to the study of commutative number fields and their arithmetics. She worked with H. Hasse and Richard Brauer, and the three collaborated on several papers concerning noncommutative algebras, the hypercomplex quantities and the theory of class fields, norm rests, and the principal genus theorem. Hasse published her theory of cross products in connection with his investigations on the theory of cyclic algebras; a paper by Brauer, Hasse, and Noether, proving that every simple algebra over an ordinary algebraic number field is cyclic, has been called a classic of its kind. As one writer put it Emmy Noether "made algebra the Eldorado of axiomatics" (Weyl 1935, p. 214).

By 1930 Emmy had established herself as the most vigorous focus for the proud mathematical tradition at Göttingen. She was an effective, innovative teacher, despite the fact that her lectures were sometimes less formal than was usual. She cared about substance, not about form or organization in her teaching; and she was stimulating, original, and quite profligate about offering others her new ideas.

Reports by students and colleagues indicate that her fertile, almost feverish imagination sparked creativity in others; her significance for algebra does not rest solely on the basis of her papers, impressive though these were. (Her name appeared on some thirty-seven publications in all.) But some of her importance to algebra has roots in her work with others, in instances where her original ideas took final shape in the work of students or collaborators. It has been said that a large part of B. L. Van der Waerden's *Modern Algebra* must be considered Emmy's contribution, and there were additional documented instances of her influence on others.

In assessing Emmy's real force, one must take note of her considerable ability to work with abstract concepts. She had the faculty of visualizing remote, very complex connections without resorting to concrete examples, and her contemporaries credit her with an unusual capacity for clarifying very difficult concepts for others.

Emmy's personal life was a quiet one during these years at Göttingen. Her days were spent at work and study in the beautiful new Mathematical Institute, which had been built at the university through the financial help of the Rockefeller Foundation. After night lectures, she would stroll home with friends through the cold, wet, dirty streets, still happily discussing complex number systems. Mathematics occupied all her hours.

Several efforts were made to secure a better position for her at the university, but prejudice and tradition continued to outweigh her scientific merits, though these were becoming more widely recognized all the time. She had been asked to deliver a course of lectures at Moscow University and another series in Frankfurt. These had brought additional recognition to her name and philosophy in the European centers of learning.

During the German revolution of 1918, Emmy had become concerned with the political and social problems of the day; although she was not actively involved in partisan politics, she gave evidence of siding with the Social Democrats. The serious struggles that shook Germany during these years helped to shape Emmy's philosophy as a pacifist, an attitude she held very strongly for the rest of her life.

Early in 1933, the rise to power of the National

Socialists sent Germany into spasms of social change from which neither the universities nor those who worked within them were exempt. Many of the academics had worn political blinders during the years leading up to the Nazis, but this apolitical stance was not enough to save them from the effects of the political upheaval that shook Germany. Emmy (along with many other scholars of whom the university had once been proud) was foreclosed from participation in any academic activity; her *venia legendi*, her appointment, and her salary were withdrawn under the constricting new dogmas of the Nazis.

Her dismissal might have been predicted, for although she had never been a political activist, she had three political strikes against her: She was an intellectual woman, a Jew, and a liberal. In the midst of the hatred, meanness, and desperate hostility rampant at the time, her stature as a mathematician could not counterbalance these three attributes.

E. T. Bell (1965, p. 261) writes of her, "She was the most creative abstract algebraist in the world. In less than a week of the new German enlightenment, Göttingen lost the liberality which Gauss cherished and which he strove all his life to maintain."

Emmy and her brother Fritz were among the lucky ones: Fritz, an applied mathematician, found refuge in the Research Institute for Mathematics and Mechanics in Tomsk, Siberia, and Emmy came to work as a professor at Bryn Mawr. She was also in demand as a lecturer at the Institute for Advanced Study in Princeton, New Jersey; and here in America, she began once more to find the hearty respect and friendship that had been denied her in Germany.

But this pleasant life, doing the work she loved among her students and colleagues, was short. After a year and a half at Bryn Mawr and Princeton, she died very suddenly on April 14, 1935, following an apparently successful operation. She was only fifty-three years old, at the apex of her productive power and technical skill. Her sturdy courage and vitality had not prepared her friends for such an early demise.

Emmy's place as a mathematician can best be judged by other mathematicians. Albert Einstein said of her (*New York Times*, May 4, 1935, p. 12):

In the judgment of the most competent living mathematicians, Fraulein Noether was the most significant creative mathematical genius thus far produced since the higher education of women began. In the realm of algebra in which the most gifted mathematicians have been busy for centuries, she discovered methods which have proved of enormous importance in the development of the present day younger generation of mathematicians.

But it remained for Hermann Weyl, an old friend who delivered her eulogy, to give her memory the warmth and affectionate vitality it deserved. In his eminently civilized tribute, there is no hint of hollow patronage; his respect for Emmy was too honest for that. His words could only have been written by someone who loved her very much (Weyl 1935, p. 219):

It was only too easy for those who met her for the first time, or had no feeling for her creative power, to consider her queer and to make fun at her expense. She was heavy of build and loud of voice, and it was often not easy for one to get the floor in competition with her. She preached mightily, and not as the scribes. She was a rough and simple soul, but her heart was in the right place. Her frankness was never offensive in the least degree. In everyday life she was most unassuming and utterly unselfish; she had a kind and friendly nature. Nevertheless she enjoyed the recognition paid her; she could answer with a bashful smile like a young girl to whom one had whispered a compliment. No one could contend that the Graces had stood by her cradle; but if we in Göttingen often chaffingly referred to her as "der Noether" (with the masculine article), it was also done with a respectful recognition of her power as a creative thinker who seemed to have broken through the barrier of sex. She possessed a rare humor and a sense of sociability; a tea in her apartments could be most pleasurable. But she was a one-sided human being who was thrown out of balance by the overweight of her mathematical talent. . . . The memory of her work in science and of her personality among her fellows will not soon pass away. She was a great mathematician, the greatest, I firmly believe, that her sex has ever produced, and a great woman.

The history of women in mathematics should not close without mention of the prodigious labors of twentieth-century women, but current history is very precarious ground, shifting and changing with disconcerting rapidity. One needs the perspective of time from which to judge the panorama that was our immediate past, and this history (as with most others) must grow sketchier and more muted as the present is approached.

The past century has witnessed a spectacular growth in the creation of pure mathematical ideas, in applied mathematics, and in the sheer number of practitioners, scholars, and teachers of mathematics. It would be impractical to attempt to deal with the histories of all the women who have rightfully earned such recognition in modern times. Their stories must be told by a scribe of another age and in another place.

But it would not be amiss, perhaps, to look at a small sample of women whose lives in many aspects parallel those of the women dealt with in earlier chapters. No critical judgment is involved in the selection of those mentioned as to the relative merit of their work. This value judgment, too, must await its proper time.

Certainly, future mathematical historians will note that the decade of the 1970s was the setting for such phenomena as the organization of the *Association for Women in Mathematics*. This organization, headed by Mary Gray, was formed to improve the status of women in the profession and to encourage more

The Golden Age
of Mathematics

women to study mathematics. In the spirit of the times, the publishers of *American Men of Science* have recently changed the name of this publication to *American Men and Women of Science*, recognizing the increase of interest and contribution among women. And future historians will surely record the illustrious career of such women as Mina Rees, who has the honor of being the first woman president of the nation's largest scientific society, the *Association for the Advancement of Science.*

The number of American women in mathematics was greatly augmented during the earlier part of the present century by European women who were interested in careers in mathematics and science and who came to America as a part of the "migration of mathematicians," a phenomenon described by Arnold Dresden (1942).

Emmy Noether was one of the women who made this migration. Lise Meitner, one of the leading mathematical physicists of this century, also made such a migration. She was born in Vienna in 1878 and grew interested in science when she read about the Curies' discovery of radium in 1902. She studied at the University of Vienna and obtained her doctorate in 1906. When she visited Germany in the following year to continue her studies, Lise had to battle the same prejudice Sonya Kovalevsky had found among German professors. Emil Fischer allowed her to work with him only after he extracted her promise that she would never enter laboratories where males were working.

Lise's work was interrupted during World War II while she served as a nurse in the Austrian Army. She was professor of physics at the University of Berlin during the twenties and was the recipient of numerous academic honors.

Because she was an Austrian, Lise was safe from the Nazi regime until the Nazi absorption of Austria in 1938. Then the purge of non-Aryans caught up with her, and Dutch scientists helped her to enter the Netherlands without a visa. With the cooperation of scientists from several countries, she found refuge in Sweden, Britain, and the United States.

Lise Meitner was strongly convinced of the actuality of uranium fission, and she published the first report concerning it from Stockholm in 1939. Her initial work helped launch procedures that eventually led to a better understanding of atomic energy, and she was given the Fermi award issued by the Atomic Energy Commission in 1966. She was the first woman to receive such an award.

Lise Meitner's devotion to science matched that of such women as Emmy Noether, Sophie Germain, and Caroline Herschel. She never married and devoted her life and time to work. She died just short of her ninetieth birthday at Cambridge, England, in 1968. Edna Kramer, in her book *The Main Stream of Mathematics* (1955, p. 193) called Lise Meitner the "foremost living successor of Kovalevsky."

Another mathematical practitioner to make the migration from Europe to America was Maria Goeppert Mayer, who studied physics, mathematics, and chemistry at Göttingen, following the tradition set earlier by Kovalevsky, Noether, and Germain, all of whom had been associated with this venerable institution. Maria Goeppert received her doctorate from the university in 1930.

She was descended from six continuous generations of German university professors. Her own career included teaching positions at Columbia University, the Institute for Nuclear Studies at the University of Chicago (where she became interested in nuclear physics), and the University of California at San Diego.

One of her best-known popular works is *Statistical Mechanics*, which she coauthored with her husband, Joseph E. Mayer, an American physicist. She was elected to the National Academy of Sciences in 1956 and was a joint winner of the 1963 Nobel Prize in physics for her work concerning nuclear shell structure. (Incidentally, Alfred Nobel's will did not provide for the prize to be awarded to mathematicians despite the enormous "benefits to mankind" deriving from their work.)

Charlotte Angas Scott was another productive mathematician who made the migration to America, though she emigrated before the turn of the century. She was educated at Girton College, Cambridge University, and her scholastic record there helped to focus public attention on the fact that women were excluded from formal admission to the "tripos" or third-year examinations and were not allowed to receive degrees from Cambridge. Ms. Scott stayed on at Girton as resident lecturer on mathematics while studying at the University of London, from which she received the doctorate of science in 1885.

She spent forty years teaching at Bryn Mawr after inaugurating the undergraduate and graduate programs in mathematics in late 1885. She published a textbook, *An Introductory Account of Certain Modern Ideas in Plane Analytical Geometry*, and she also authored some thirty papers published in mathematical journals in several countries. Her specific interests centered on the analysis of singularities for algebraic curves. She was also active in mathematical organizations and societies before she retired at the age of sixty-seven and returned to England.

Not all the women who left Europe during the migration of mathematicians came to America. Hanna von Caemmerer, an algebraist, left her native Germany in 1938 for England. Her fiancé, Bernhard Neumann, also an abstract algebraist had emigrated several years earlier. Hanna was an Aryan, but her fiancé was not, and Nazi dogma did not allow the marriage of the two.

She continued her research in specialized mathematical fields and received her doctorate at Oxford. (Incidentally, she too had studied earlier at Göttingen.) Her major interest was in group theory (specifically in free groups), and she was a lucid and prolific writer, rivaling Emmy Noether in her clear literary style. Her numerous research memoirs deal with such topics as finite nondesarguesian planes, discussed in terms that are within the understanding of the general reader. Ms. Neumann died late in 1971, and plans are currently underway to endow a prize in pure mathematics at the Australian National University in her honor.

It is of interest to note that the tradition of learned Italian women is being kept alive by such mathematicians as Maria Pastori, who attended the Maria Agnesi School and later became professor at the Istituto Matematico of the University of Milan. Her work centers on the fields of tensor analysis and relativity. Unlike the Agnesi family, the Pastoris were not wealthy, and Maria's early education was the result of her own initiative and hard work. Edna Kramer (1957, p. 86) calls her the "true daughter of Italy and a twentieth-century disciple of Agnesi." Maria Pastori's sister was responsible for having a nursing school named for Agnesi in memory of her services to the Milanese poor.

During the two centuries following Agnesi's time, geometry experienced a powerful development, particularly algebraic and differential geometry. The tensor calculus, of central importance in the theoretical treatment of classical physics, is useful today in the pure mathematical investigation of generalized spaces. Maria Pastori's work has helped to extend the usefulness of the tools needed for these investigations.

Maria Cibrario has also continued the brilliant tradition of Italian women in mathematics. She filled the chair of mathematical analysis first at the University at Modena, then the corresponding chair at Pavia. Her work and research have helped to achieve the classification of linear partial differential equations of the second order of mixed type (including for many of these existence and uniqueness theorems). She has also researched nonlinear hyperbolic equations and systems of such equations, and she is credited with solving the Goursat problem for the hyperbolic nonlinear equation of the second order. Her work in these branches of analysis has gone far beyond that of her predecessor in this field, Sonya Kovalevsky.

Among twentieth-century French mathematicians, the name of Jacqueline Lelong-Ferraud, professor of mathematics at the University of Paris, is notable. She continued the tradition of Emilie du Châtelet and was among the first females to take the examination for entry into the famous Paris Ecole Normale Supérieure. She went on to full professorship in pure mathematics at the University of Paris. Her work includes research on the behavior of conformal transformations and representations, Riemann manifolds and harmonic forms, potential theory, and so forth. She is credited with originating the concept of preholomorphic functions, using these to produce a new methodology for proofs.

In the field of algebraic topology, the name of Paulette Liberman, professor at the University of Rennes, should be noted. Although World War II seriously interrupted her professional career (Vichy laws excluded her from effective work until after the Liberation), this remarkable mathematician went on to become a member of Charles Ehresmann's research school and to work on differentiable fiber spaces, almost complex manifolds, and their generalizations. She was a protégé and student of Elie Cartan, one of the greatest geometers the twentieth century has produced. She also studied at Oxford with A. N. Whitehead, a disciple of Cartan's.

In Russia, Sophie Kovalevsky had her counterpart among modern mathematicians in the person of another Sophie. Sophie Picard was born at St. Petersburg and educated at the University of Smolensk, where her father was a professor of natural sciences. She was surrounded by a literate and intellectual family, as was the earlier Sophie, and both her parents helped her to prepare for a scientific career.

The Picards left postrevolutionary Russia in the 1920s and migrated to Switzerland, where Sophie studied and earned her doctorate at the University of Lausanne. Her father's death and the consequent financial problems forced her to take a job as an actuary, but her free hours were spent in study and research. She was eventually able to move into an academic position of distinction and came in time to occupy the chair of higher geometry and probability theory at the University of Neuchâtel. The modern student of statistics and probability theory encounters her name frequently in discussions of group theory, function theory, the theory of relations, and so on.

And future historians will find cause to deal with such contemporary mathematicians as Olga Taussky Todd, who was trained in number theory and who worked with David Hilbert at Göttingen. Along with other members of the Hilbert school who emigrated to America, she has enriched American mathematics immeasurably.

Emma Lermer's work with special cases of Fermat's last theorem will surely be noted, as will Julia Robinson's contribution to Hilbert's 10th problem. The names of Elizabeth Scott, Grace Hopper, and Dorothy Maharam Stone, among scores of others, will also appear in the mathematical histories of tomorrow.

It would be easy from the foregoing brief look at women in modern mathematics to aphorize that during the past few decades, the prospect of equality for women in mathematics has improved. Such is not the case; indeed, there are many indications to the contrary.

A recent report by the National Research Council

states (1968, p. 50): "Many mathematicians believe that this is a golden age of mathematics." In a sense this is true, for mathematicians are now helping to solve many of the baffling problems confronting humankind. It may even be true for many women in mathematics, but certainly it is not true for all; and although the number of women in mathematics is increasing rapidly, their proportion of the total number of mathematicians is declining as men are being trained at a far greater rate than women. In terms of education, occupation, and income, as the level increases, women represent a smaller proportion. It is, then, discouraging but true that since the early part of the twentieth century the position of women in mathematics has been steadily worsening rather than improving.

In his fantasy, *Through the Looking Glass*, Lewis Carroll has the Red Queen say to Alice, "It takes all the running you can do to keep in the same place. If you want to get somewhere else, you must run at least twice as fast as that!" (p. 1189).

Carroll (or C. L. Dodgson) was an English mathematician and logician who drew on his specialized knowledge to perfect the art of imaginative writing. It is surely a part of his genius that he has this dialogue take place between two females, for nowhere is this metaphor more applicable than it is for women in mathematics.

In almost any age, it has taken a passionate determination, as well as a certain insouciance, for a female to circumvent the crippling prohibitions against education for women, particularly in a field that is considered to be a male province. In mathematics, the wonder is not that so few have attained proficiency in the field, but that so many have overcome the obstacles to doing so. We can only speculate about the multitude who were dissuaded from the attempt—the Mary Somervilles who never had a fortunate accident to discover their talent, the Agnesis who lacked a mathematically trained parent to nurture their genius, or the Mme du Châtelets who were seduced completely by a frivolous salon life.

But perhaps the larger tragedy is that, even today, we can find remnants of the elitist (or sexist) tradition that has so often surrounded mathematics in the past. It should be acknowledged that during the present

The Feminine Mathtique

century, there have been many women who have achieved remarkably successful careers in fields drawing heavily on mathematics, but to use these women as exemplars of what is possible for any woman who "really tries" is one of the crueler sports of our day. That so many of the resolute *do* survive speaks to their capabilities and circumstances, as well as the caprice of luck and nature. Far too many fail even to see the reasons they were dissuaded from the effort.

By glamorizing the exceptional case, we often manage to preserve the myth of equality in education and convince ourselves that when women fail to become educated in such disciplines as mathematics or the "hard sciences," the lack of individual enterprise or interest was the important factor, rather than the lack of equal opportunity. Fortunately, we are beginning to have a more sophisticated and sympathetic understanding of the socialization process that shapes the female experience.

The mathematization of the physical sciences has gone on for centuries, but since World War II there has been an accelerated mathematization of other aspects of our culture. This permeation of the mathematical method into an increasing variety of human endeavors and studies coincides roughly with the development in our society of what Betty Friedan has called the "feminine mystique," which also grew strongly during this same period of time.

These two phenomena, juxtaposed as they are, have

combined to foreclose many women from meaningful participation in intellectual and economic activities and have also produced a complex of attitudes both in and about women that might be called, for want of a better term, the *feminine mathtique.*

This mathtique has served to perpetuate the destructive and pervasive myths concerning women's aptitudes, accomplishments, and ambitions in mathematical endeavors. It encourages the notion that to enjoy mathematics in its many forms is to be, in some obscure way, at variance with one's womanhood. It also perpetuates the socialization process that reinforces and promotes this assumption. It breeds and institutionalizes graceless jokes and stereotypes about the helpless, checkbook-bumbling female, the mindless housewife, the empty-headed husband-chasing coed, the intuitive (but illogical) woman who "hates arithmetic." (For many of the women mentioned in these biographies, it was the early support of an intelligent and mathematically educated father that made the crucial difference in the socialization process.)

The mathtique finds expression not only in the deliberate bias of many "important others" in a young girl's life but also in myriad more covert media, such as the lack of successful role models, improper teaching and counseling approaches, strong social emphasis on other interests and life-styles, and even in such small ways as the lack of feminine content in textbook problems, which has been shown to have its negative effect also. (It is interesting to note that over the past decade, Sweden and Denmark have been moving to reduce the masculine content of mathematics textbooks.)

This kind of psychological conditioning produces expectations within and about the young female student concerning her ability and performance in those male-dominated fields that make a more rigorous use of mathematics. It is difficult indeed to protect the young girl growing up in our society from contamination by this set of expectations, for they are to be found too often within the formal educational process itself.

Nor is this the end of it. These attitudes reach beyond the educational process and into the economic and professional life of women workers, no doubt accounting in part for the salary differential between male and female mathematicians. (The median annual salary for female "full-time civilian scientists" is $9,400; for males it is $13,000 according to statistics compiled by the Women's Bureau of the Labor Department in 1969 [National Science Foundation 1968].)

Concern has been expressed in many quarters over this waste of intellectual resources and the ineffective use of the potential of women. Those who are charged with the development of the mathematical sciences are concerned about both the loss of those women who have the potential but because of psychological blocks never become trained and the loss of highly trained women mathematicians.

Recognition of some of the problems women face in the formal study of both applied and pure mathematics may be found in volumes published by the National Academy of Sciences; in reports by the Committee on Support of Research in the Mathematical Sciences; and in reports by the Panel on Undergraduate Education in Mathematics. Recognition of these problems was also manifested in congressional hearings held recently before the Special Sub-committee on Education, chaired by Edith Green of Oregon.

As testimony before Representative Green's committee revealed, academic quota systems still exist at many universities and colleges, whether openly admitted or not. Testimony also indicated that the consistently low percentage of women students in certain fields of study might be explained by the higher standards of admission for women at all levels, denial of loans, assistantships, and fellowships, deterrence from part-time study, course scheduling, and other policies and practices geared to the service of male students and instructors. Among the more subtle forms of bias with which women must cope is the pressure on female students to choose traditionally popular fields of study, the socially approved lower self-aspiration for women, and, perhaps most important of all, the expectations of others.

It is interesting to note that with the increased emphasis on mathematics and science after the

Sputnik panic, it was anticipated that more girls would be attracted to these fields. Findings by Poffenberger and Norton (1963, pp. 341–350) indicate that this has not been the case and that the resulting curricula changes have appealed more to males than to females. It is a problem that was not caused by educators alone, and educators by themselves cannot solve it. But if the ideological goal of achieving true equality is to be realized, then women and interested educators must discuss and define concrete programs aimed at effecting change at all levels of education, attracting more women to this field of study, researching and understanding the psychological dynamics underlying the difficulties women encounter in this field, and offering compensatory training where necessary.

In our statisticized and computerized world, mathematical illiteracy effectively blocks one not only from an understanding of the physical sciences but also from substantial portions of current literature in many other disciplines. As Gelbaum and March (1969) point out in the preface to their book:

It was once appropriate, and still is rather conventional, for a parent or counselor to advise a student who finds mathematics uncongenial to consider studying the social and behavioral sciences. The advice has the virtue of providing a neat solution to a difficult problem; it has the vice of being misguided.

Within the past two decades, mathematics has become indispensable to the student of human behavior.

This incursion into the social and behavioral sciences is not an exception to the rule: very nearly all academic disciplines are rapidly moving toward an increased use of research premises, design methodology, and interpretive analysis, all of which exact more mathematical sophistication.

In addition, the new and emerging relational disciplines not only draw on traditional mathematics but in some cases have served as moments for the creation of specialized branches of mathematics. There are ever-growing demands within our educational system for increased mathematical insight; as matters stand, each female student must work out individual solutions to the formidable problems of acquiring the needed mastery.

It is the immodest aim of this volume to annul a small part of the feminine mathtique by marshaling relevant information about the women who have left us such a rich legacy in mathematics and by imparting a modicum of understanding for their age-long struggle. Their efforts, whether born of a sense of purpose, simple indulgence, or total self-belief, seemed to spring naturally from the life of each individual; contemporary mathematics would only have been diminished without them and their work.

In his autobiography, Bertrand Russell listed the desire to "apprehend the Pythagorean power by which number holds sway above the flux" as one of the three governing passions of his life. A fortunate constellation of circumstances encouraged Lord Russell in the pursuit of this passionate life task. But one is led to speculate that, had the single parameter of sex been changed, this passion and much of its ultimate reward might have been lost to the world.

Hypatia

From Elbert Hubbard, *Little Journeys to the Homes of Great Teachers* (New York: The Roycrofters, 1908).

Maria Agnesi

From M. Jacques Boyer, "Sketch of Maria Agnesi," *Popular Science Monthly*, volume 53 (New York: D. Appleton & Company, July 1898), p. 289.

Emilie de Breteuil, Marquise du Châtelet

Portrait by Nicolas de Largillière, courtesy of The Bettman Archive, Inc.

Caroline Herschel

From James Parton, *Noted Women of Europe and America* (Hartford, Connecticut: Phoenix Publishing Company, 1883).

Sophie Germain

From the *Grand Larousse encyclopédique*, volume 5 (Paris: Librairie Larousse, 1962), p. 459.

Mary Fairfax Somerville

Courtesy of The National Gallery of Scotland, Scottish National Portrait Gallery of Scottish National Gallery of Modern Art.

Sonya Corvin-Krukovsky Kovalevsky

Courtesy of Stockholms Universitet.

Emmy (Amalie) Noether

Courtesy of *Scripta Mathematica*, volume 3, 1935.

Illustration Credits

Adler, Alfred, 1972. "Mathematics and Creativity." *The New Yorker*, 19 February 1972.

Agnesi, Maria Gaetana, 1801. *Analytical Institutions. Translated by John Colson. London: Taylor and Wilks.*

American Men and Women of Science, 1972. Twelfth Ed., Jacques Cattell Press, ed. New York: R. R. Bowker Company.

Beard, Mary R., 1931. *On Understanding Women.* New York: Longmans, Green & Company.

Beard, Mary R., 1946. *Women as a Force in History.* New York: The Macmillan Company.

Bell, E. T., 1945. *Development of Mathematics.* New York: McGraw-Hill Book Company.

Bell, E. T., 1951. *Mathematics, Queen and Servant of Science.* New York: McGraw-Hill Book Company.

Bell, E. T., 1937 (1961, 1965). *Men of Mathematics.* New York: Simon and Schuster.

Boyer, M. Jacques, 1898. "Sketch of Maria Agnesi." *Popular Science Monthly* 53 (July 1898). New York: D. Appleton & Company.

Coolidge, Julian L., 1951. "Six Female Mathematicians." *Scripta Mathematica* 17 (March–June).

Cornford, F. M., 1953. *The Cambridge Ancient History.* Vol. 6, J. B. Bury et al., eds. London: Cambridge University Press.

Dictionary of National Biography, 1922. Vol. 18. London: Oxford University Press.

Discrimination Against Women, 1970. Hearings before the Special Subcommittee on Education of the Committee on Education and

References

Labor, House of Representatives. Section 805 of H. R. 16098. Washington: U.S. Government Printing Office.

Dresden, Arnold, 1942. "The Migration of Mathematicians." *American Mathematical Monthly* 49.

Einstein, Albert, 1935. "The Late Emmy Noether." *The New York Times*, 4 May 1935.

Encyclopedia Britannica, 1966. Vol. 14. Chicago: William Benton, Publisher.

Gamble, Eliza B., 1916. *The Sexes in Science and History.* New York: G. P. Putnam's Sons.

Gelbaum, Bernard R., and James G. March, 1969. *Mathematics for the Social and Behavioral Sciences: Probability, Calculus and Statistics.* Philadelphia: W. B. Saunders Company.

Gibbon, Edward, 1960. *The Decline and Fall of the Roman Empire.* (An abridgement by D. M. Low). New York: Harcourt, Brace, and World.

Grand Larousse encyclopédique, volume 5, 1962. Paris: Librairie Larousse.

Hale, Sarah Josepha, 1860. *Woman's Record: or Sketches of All Distinguished Women from the Creation to A. D. 1854.* New York: Harper & Brothers Publishers.

Hamel, Frank, 1910. *An Eighteenth-Century Marquise.* London: Stanley Paul & Company.

Heath, Thomas L., 1964. *Diophantus of Alexandria: A Study in the History of Greek Algebra.* New York: Dover Publications.

Herschel, Mrs. John, 1876. *Memoir and Correspondence of Caroline Herschel.* London: John Murray.

Hubbard, Elbert, 1908. *Little Journeys to the Homes of Great Teachers*. Vol. 23. New York: The Roycrofters.

Jaffe, Bernard, 1944. *Men of Science in America*. New York: Simon & Schuster.

Jowett, Benjamin, trans., 1892. *Plato's Dialogues*. London: Clarendon Press.

Kennedy, Hubert C., 1969. "The Witch of Agnesi—Exorcised." *The Mathematics Teacher*, October 1969.

Kingsley, Charles, 1853. *Hypatia or New Foes with Old Faces*. Chicago: W. B. Conkley Company.

Knapp, R. H., and H. B. Goodrich, 1952. *Origins of American Scientists*. Chicago: The University of Chicago Press.

Kramer, Edna E., 1955. *The Main Stream of Mathematics*. New York: Oxford University Press.

Kramer, Edna E., 1957. "Six More Female Mathematicians." *Scripta Mathematica* 23.

Kramer, Edna E., 1970. *The Nature and Growth of Modern Mathematics*. New York: Hawthorn Books, Inc.

Leffler, Anna Carlotta, 1895. *Sonya Kovalevsky, Her Recollections of Childhood, with a Biography*. New York: The Century Company.

London Society, An Illustrated Magazine of Light and Amusing Literature for Hours of Relaxation, 1870. Vol. 28. London: Wm. Clowes and Son.

Maccoby, Eleanor, 1966. *The Development of Sex Differences*. Stanford, California: Stanford University Press.

Meschkowski, Herbert, 1964. *Ways of Thought of Great Mathematicians*. San Francisco: Holden-Day.

Milton, G. A., 1957. "The Effects of Sex-Role Identification upon Problem-Solving Skill." *Journal of Abnormal Psychology* 55.

Mitford, Nancy, 1957. *Voltaire in Love*. London: Hamish Hamilton.

Mittag-Leffler, G., 1893. "Sophie Kovalevsky." *Acta Mathematica* 16 (1892–1893).

Mozans, H. J., 1913. *Woman in Science*. New York: D. Appleton and Company.

National Research Council, 1968. *The Mathematical Sciences: Undergraduate Education*. Washington, D.C.: National Academy of Sciences.

National Science Foundation, 1968. "National Register of Scientific and Technical Personnel." Reported in U.S. Department of Labor, WB 71–86. Washington, D.C.: U.S. Government Printing Office, 1971, 0–419–373.

Osen, Lynn M., 1971. *The Feminine Math-tique*. Pittsburgh, Pa.: KNOW, INC.

Parton, James, 1883. *Noted Women of Europe and America*. Hartford, Connecticut: Phoenix Publishing Company.

Poffenberger, T., and D. Norton, 1963. "Sex Differences in Achievement Motive in Mathematics as Related to Cultural Change." *Journal of Genetic Psychology* 103.

Proctor, Richard A., 1886. *Light Science for Leisure Hours*. Second series. London: Longmans, Green and Company.

Reid, Constance, 1970. *Hilbert*. New York: Springer-Verlag.

Rektorys, Karel, ed., 1969. *Survey of Applicable Mathematics.* Cambridge, Mass.: MIT Press.

Richeson, A. W., 1940. "Hypatia of Alexandria." *National Mathematics Magazine*, November 1940.

Richeson, A. W., 1941. "Mary Somerville." *Scripta Mathematica* 8, No. 1 (March 1941).

Seltman, Charles, 1956. *Women in Antiquity.* London: Pan Books, Ltd.

Smith, David Eugene, 1951. *History of Mathematics.* Vol. 1. Boston: Ginn and Company.

Starr, Chester G., 1971. *The Ancient Greeks.* New York: Oxford University Press.

Struik, D. J., 1969. *A Source Book in Mathematics 1200–1800.* Cambridge, Mass.: Harvard University Press.

Sullivan, J. W. N., 1925. *The History of Mathematics in Europe.* London: Oxford University Press.

Tabor, Margaret E., 1933. *Pioneer Women.* London: The Sheldon Press.

Taylor, Eva G. R., 1966. *Mathematical Practitioners of Hanoverian England.* London: Cambridge University Press.

Thomas a Kempis, Sister Mary, 1940. "The Walking Polyglot." *Scripta Mathematica* 6.

Thomas a Kempis, Sister Mary, 1955. "Caroline Herschel." *Scripta Mathematica* 1 (June).

Todhunter, Isaac, and Karl Pearson, 1960. *A History of the*

Theory of Elasticity and of the Strength of Materials. New York: Dover Publications.

Turnbull, Herbert W., 1961. *The Great Mathematicians.* New York: New York University Press.

Wade, Ira O., 1941. *Voltaire and Madame du Châtelet: An Essay on the Intellectual Activity at Cirey.* Princeton: Princeton University Press.

Wade, Ira O., 1969. *The Intellectual Development of Voltaire.* Princeton: Princeton University Press.

Weyl, Hermann, 1935. "Emmy Noether." *Scripta Mathematica* 3.

Whitney, Charles A., 1971. *The Discovery of Our Galaxy.* New York: Alfred A. Knopf.

"Women in Mathematics," 1970. *The Arithmetic Teacher*, April 1970.

Woodruff, L. S., ed., 1923. *Development of the Sciences.* New Haven: Yale University Press.

World Who's Who in Science: A Biographical Index of Notable Scientists from Antiquity to the Present, 1968. Allen G. Debus, ed. Chicago. A. N. Marquis Company.

Index